"THE BUCK STOPS HERE":

Senior Military Commanders on Operations

IN HARM'S WAY

"The Buck Stops Here":
Senior Military Commanders
on Operations

Edited by
Colonel Bernd Horn

CANADIAN DEFENCE ACADEMY PRESS

Copyright © 2007 Her Majesty the Queen, as represented by the Minister of National Defence.

Canadian Defence Academy Press
PO Box 17000 Stn Forces
Kingston, Ontario K7K 7B4

Produced for the Canadian Defence Academy Press
by 17 Wing Winnipeg Publishing Office.
WPO30265

Cover Photo:
Master Corporal Brian Walsh for DGPA/J5PA Combat Camera

Library and Archives Canada Cataloguing in Publication

The buck stops here : senior military commmanders on operations / edited by Bernd Horn.

(In harm's way)
Issued by: Canadian Defence Academy.
Includes bibliographical references and index.
ISBN 978-0-662-45813-5 (bound).-- ISBN 978-0-662-45814-2 (pbk.)
Cat. no.: D4-4/3-1-2007E (bound) -- Cat. no.: D4-4/3-2-2007E (pbk.)

1. Canada--Armed Forces--Foreign countries. 2. Canada--Armed Forces--History--20th century. 3. Canada--Armed Forces--Officers. 4. Military art and science--Canada. 5. Command of troops. 6. Operational art (Military science). I. Horn, Bernd, 1959- II. Canadian Defence Academy III. Title. IV. Title: Senior military commmanders on operations. V. Series.

UA600.B82 2007   355.30971'09049   C2007-980107-2

Printed in Canada.

1 3 5 7 9 10 8 6 4 2

ACKNOWLEDGEMENTS

I would be remiss if I did not acknowledge the hard work and efforts of all those who made this volume possible. First, I would like to thank the contributors for their dedication and sense of duty. Although it is the obligation of all leaders to prepare their successors for the challenges that may be faced in the future, not all rise to the task. I fully realize that the contributions within this book were the efforts of off-duty hours – nights, weekends and holidays. Their efforts are greatly appreciated.

Equally important, I wish to thank the Lessons Learned Cell at CFLI, specifically Commander Bob Edwards, Major Brent Beardsley and Craig Mantle for their diligent campaign at identifying and recruiting contributors for this book. Their research and aggressive campaign to convince individuals to participate was seminal to the success of the "In Harm's Way" project. I must also extend my sincere appreciation to Joanne Simms and Carol Jackson for their continuing assistance in the administration and translation of the these projects.

Finally, I must pass my sincere gratitude to the technical component of CDA Press, namely 17 Wing Publishing Office - specifically Captain Phil Dawes, Evelyn Falk, Mike Bodnar and Adrienne Popke, for their expertise and professionalism in turning the raw manuscript into a refined, finished publication.

IN HARM'S WAY

TABLE OF CONTENTS

Foreword iii

Introduction v

Chapter 1   Commanding UN Military Observers in Sector Sarajevo 1993-1994
Roy Thomas 1

Chapter 2   Operation TOUCAN Overview
Rear-Admiral Roger Girouard 27

Chapter 3   In Harm's Way: The 3rd Battalion, The Royal Canadian Regiment on Operation PALLADIUM, ROTO 3 – July 1998 to January 1999
Colonel Mike Jorgensen 47

Chapter 4   Operation PALLADIUM ROTO 13, September 2003-April 2004
Colonel Dean Milner 83

Chapter 5   Command in Combat
Brigadier-General D.A. Davies 89

Chapter 6   Operation APOLLO
Lieutenant-Colonel G.L. Smith 109

Chapter 7   Operation HALO: A Leadership Challenge
Lieutenant-Colonel Pierre St-Cyr 137

Chapter 8   Observations on UN Strategic and Institutional Leadership – South Sudan 2005-2006.
Lieutenant-Colonel Mike Goodspeed 163

Chapter 9   Civil-Military Operations in Combined Force Command – Afghanistan
Colonel François Vertefeuille 179

| Chapter 10 | Full Spectrum Leadership Challenges in Afghanistan
Colonel Bernd Horn . . . . .191 |
|---|---|
| Chapter 11 | Reflections on Afghanistan:
Commanding Task Force Orion
Lieutenant-Colonel Ian Hope . . . . .211 |
| Chapter 12 | In the Breach:
The Combat Command of Lieutenant-Colonel Omer Lavoie
Colonel Bernd Horn . . . . .227 |
| Chapter 13 | No Small Victory:
Insights of the Commander of Combined Task Force Aegis on Operation MEDUSA
Brigadier-General David Fraser . . . . .243 |
| Chapter 14 | We Three Hundred:
Logistics Success in the New Security Environment
Lieutenant-Colonel John Conrad . . . . .257 |

Contributors 271

Glossary of Abbreviations and Acronyms 275

Index 281

FOREWORD

I am delighted to introduce *"The Buck Stops Here": Senior Military Leadership on Operations*. This book represents the third volume of the seminal *In Harm's Way* series created by the Canadian Forces Leadership Institute (CFLI) and the twenty-third book published by the CDA Press. As such, the Canadian Defence Academy is well on its way to capturing operational experience that can be used in our professional development institutions. After all, this book, like those before it and those still in press, are an integral component of our Strategic Leadership Writing Project, which is designed to (a) create a distinct and unique body of Canadian leadership literature and knowledge that will assist leaders at all levels of the Canadian Forces in preparing themselves for operations in a complex security environment, and (b) inform the public with respect to the contribution of Canadian Forces service personnel to Canadian society and international affairs.

This volume, as well as those that preceded it and those that will follow it in the *In Harm's Way* series, represents the experiences and insights of an array of individuals who have taken the time and effort to capture their thoughts so that others may profit from them. Invariably, this means utilizing personal leave, weekends and nights to complete their stories. Their commitment and dedication is greatly appreciated. I wish to thank all those who have risen to the challenge of preparing our successors for the complexities that lie ahead.

"The Buck Stops Here" is an important addition to the series. Similar to the other volumes, it contains a collection of "war stories" from all three environments and from a myriad of different operations. They are based on personal experiences and the respective interpretations, reflections and lessons learned. Although these experiences do not represent the doctrine and policy of the Department of National Defence or of the Canadian Forces, they are no less valid. They express a richness of information that can assist others in preparing for operations and leadership in general. In essence, they can act as virtual experience for those who have not had the opportunity to deploy. However, even those with operational experience can profit from a wider, broader and greater repertoire of knowledge.

In closing, I wish to reiterate the importance of this book, as well as that of all others in the *In Harm's Way* series. At the Canadian Defence Academy, we hope that they will provide valuable insight for those who serve in, and for those who interact with, the profession of arms in Canada.

Major-General P.R. Hussey
Commander, Canadian Defence Academy

INTRODUCTION

Senior Military Leadership on Operations is the third volume of the Canadian Forces Leadership Institute's (CFLI) *In Harm's Way* series. Consistent with the first two volumes, its strength lies in the experience of the contributors. Again, I feel obliged to highlight the dedication, initiative and professionalism of the men and women in the Canadian Forces (CF) who unselfishly give of themselves to serve their nation and the people of Canada. Quite simply, Canadian military personnel never fail to amaze me. Consistently they have done their country proud, persevering under austere, harsh and remote conditions, as well as in dangerous and volatile environments. They have faced ambiguity, change and uncertainty with courage and conviction. And despite constraints in manpower and equipment, they have never failed the people or the government of Canada. Moreover, they have always been humble, if not embarrassingly discreet, about their achievements.

Although this culture of understatement and humility is exceptionally commendable, it does have a significant weakness. First, such culture fails to educate Canadian society in regard to the contribution and sacrifice of its military personnel. As such, Canadians, including politicians, are not always fully aware of the difficult nature or the significance of CF personnel's achievements.

More important, the reluctance to share experiences deprives the institution of valuable learning opportunities. Although the security environment is constantly changing and each deployment or theatre of operation offers its own unique challenges, at least knowledge, if not vicarious experience, can be derived from the actual experience of others. Leadership challenges in particular offer a bonanza of professional development opportunities. There are many situations, problems and dilemmas that are timeless and transcend missions or geographic area. The sharing of these challenges and different approaches and/or solutions by individuals provides guidance to others, especially young, inexperienced leaders. Similarly, lessons derived from specific missions, operations or geographic areas also provide the insight and knowledge to those deploying that will allow them to be better prepared to meet the challenges and to lead their personnel more effectively.

The chapters also provide a window to the different cultures of the various services (Navy, Army and Air) as well as on the myriad of problems that confront individuals in the various classifications and trades of the military occupation. This window will assist all military personnel, as well as the public at large, with gaining a clearer understanding of the peculiar and distinct challenges that individuals face. Such comprehension is critical as we move forward towards a more integrated approach in operations.

Overall, this sharing of knowledge and experience throughout the institution is key to enabling mission success. It is also the cornerstone of a learning organization. For this reason, CFLI established the Strategic Leadership Writing Project, a seminal program intended to create a distinct and unique body of Canadian leadership literature and knowledge that will assist leaders at all CF rank levels to prepare themselves for operations in a complex security environment. The project will also educate the public in regard to the contribution of CF service personnel to Canadian society and international affairs.

The *In Harm's Way* series encapsulates "war stories" (for lack of a better term) that capture the challenges of leaders at all rank levels and from all three services, in operations ranging from the post-Cold War period to the early 1990s to the present. These stories are intended to act as a professional development tool for military members, as well as an educational vehicle for the Canadian public.

This volume, as the title suggests, focuses on the command and leadership experiences of senior officers on operations. As such, it is important to lay down some fundamental markers on the concept of command and leadership. Command is a very personal experience. How an individual commands speaks more of the character and personality of the respective person than it does to the concept of command itself. In essence, command is far more an art than it is a science. That is why there is such a wide variance between commanders – some who reach legendary status and others who fade into ignominy.

Command, however, is not an arbitrary activity. It can only be exercised by those who are formally appointed to positions of command. It is "the authority vested in an individual of the armed forces for the direction, co-ordination, and control of military forces."[1] And this normally comes after years of formal professional development, assessment and evaluation, as well as proven ability.

The necessity to prepare individuals for command is not surprising. First of course is the heavy responsibility that command entails – namely, the lives of others. Commanders must accomplish their tasks but they must do so with the minimum cost in lives and within their allocated resource envelope. This is often an enormous challenge. The all encompassing scope of command is why it comprises of three, often reinforcing, components. They are: authority, management (e.g. allocating resources, budgeting, coordinating, controlling, organizing, planning, prioritising, problem solving, supervising and ensuring adherence to policy and timelines) and leadership (i.e. "directing, motivating and enabling others to accomplish the mission professionally and ethically, while developing or improving capabilities that contribute to mission success").[2]

Depending on the mission, subordinates, circumstances and situations, as well as the respective commander, different emphasis is placed on the different components. It must be noted that command can only be exercised by those who are appointed to such positions. Conversely, leadership, which is a component of command, but also exists outside of the concept of command, can be exercised by anyone.[3] The fact that command is a function of appointment and leadership is a result of a voluntary interaction between someone practicing leadership and those accepting to be led, is one of the greatest defining difference between the concepts of command and leadership.

With that in mind, similar to the first two volumes, the chapters contained in *Senior Leadership on Operations*, are the experiences and viewpoints of the authors and should be taken as such. Personal attitudes, biases, beliefs and interpretation are at play. Readers may not agree with all the views, opinions or statements made. This is to be expected. What is contained in this book, or in any volume of the series, is not meant to represent CF doctrine or official policy. Rather, it is the personal experiences and reflections of individuals who have been in harm's way in the service of their country. Their narratives are offered in the spirit of critical thought and shared experiences, in an effort to help others to be better prepared when they are called out on operations. In addition, the lessons that leap from the pages should serve to generate discussion and debate in order to determine better practices and policies that will serve to make the CF more effective, efficient and operationally ready. Furthermore, the contributions should act as a means of sharing with the Canadian people glimpses of the challenges and hardships that their military endure and the accomplishments that their military achieve.

ENDNOTE

1   Canada, *Command* (Ottawa: DND, 1997), 4. The essence of command is the expression of human will – an idea that is captured in the concept of commander's intent as part of the philosophy of mission command. The Commander's intent is the commander's personal expression of why an operation is being conducted and what he hopes to achieve. It is a clear and concise statement of the desired end-state and acceptable risk. Its strength is the fact that it allows subordinates to exercise initiative in the absence of orders, or when unexpected opportunities arise, or when the original concept of operations no longer applies. Mission Command is a command philosophy that promotes decentralized decision-making, freedom and speed of action and initiative. It entails three enduring tenets: the importance of understanding a superior commander's intent, a clear responsibility to fulfil that intent, and timely decision-making. In sum, command is the purposeful exercise of authority over structures, resources, people and activities.

2   Canada, *Leadership in the Canadian Forces: Conceptual Foundations* (Kingston: DND, 2005). It is within this powerful realm of influence and potential change that leadership best demonstrates the fundamental difference between it and the concept of command. Too often, the terms leadership and command are interchanged or seen as synonymous. But, they are not. Leadership can, and should, be a component of command. After all, to be an effective commander the formal authority that comes with rank and position must be reinforced and supplemented with personal qualities and skills – the human side. Nonetheless, as discussed earlier, command is based on vested authority and assigned position and / or rank. It may only be exercised downward in the chain of command through the structures and processes of control. Conversely, leadership is not constrained by the limits of formal authority. Individuals anywhere in the chain of command may, given the ability and motivation, influence peers and even superiors. This clearly differentiates leadership from command.

3   Command and leadership are often incorrectly intermixed. Each is a distinct concept. A commander should use leadership, but technically does not have to. He or she can rely exclusively on authority and take a very managerial approach (but this is not to say that managers do not use leadership – like commanders they certainly should).

CHAPTER 1

COMMANDING UN MILITARY OBSERVERS IN SECTOR SARAJEVO, 1993-1994

Roy Thomas

The intended tour to facilitate the handover between the United Nations Military Observers (UNMO) in Sector Sarajevo, specifically to show the various Observation Posts (OPs) and team sites was not going well. It had been cut short while we were inspecting the "hotspot" on Mount Igman.[1] Ongoing fighting meant we would have to take the short cut to Hadicizi, which would require us to offer ourselves as potential targets to some anti-aircraft gunners while racing down hill at best Totoya Landcruiser speed on a spiralling open road. "We'll floor it when we break out of the trees," the observer driving the vehicle told me. As I had just come from driving myself on the tracks of the Macedonian/Albanian border, entrusting my life to someone else's skill as a chauffer was nerve-wracking. Fortunately, for the first time in days, our speedy descent was not punctuated by cannon fire. That was the good news of my first day in Sector Sarajevo!

The bad news was that an UNMO team that was being evacuated by a French armoured personnel carrier (APC) from an UNMO OP that was under almost constant Bosnian Serb shell fire had just been stopped by a Bosnian Army checkpoint. It got worse! So-called Bosnian Army personnel had hi-jacked the APC itself and all the contents including personal weapons and wallets of the UN Protection Force UNPROFOR) personnel inside. The UNMOs of course were unarmed. A French APC had been used because although the UNMOs in Sarajevo had some armoured General Motors Corporation (GMC) suburbans, these vehicles were designed to protect VIPs (very important persons) against snipers not artillery fire, therefore, they had no armoured roof.

This was my introduction to Sector Sarajevo. The day that I arrived to assume command of UNMO Sarajevo, the teams were spread out as follows: "P" (papa) team was in OPs inside the city; "L" (lima) team was scattered among the Bosnian Serbs surrounding the Bosnian capital, with one group in the so-called safe haven of Gorazde and another group in the pocket of Zepa, both in Eastern Bosnia on the banks of the Drina. No wonder the French Sector Commander, General Soubirou, was not in a good mood when I made my courtesy call![2]

"Welcome to Hell," said the graffiti smeared on the wall of a building! "Welcome" indeed to sharing some detail of my nine months as the senior UN military observer for Sector Sarajevo, 15 October 1993 to 15 July 1994, commanding anywhere from 120 to 200 officers from any of 39 countries from all continents. In reality, it was not hell for me, but rather the most challenging and rewarding assignment of my 35 years service in the Canadian Forces (CF).

In the following pages, I wish to outline what worked for me and my command of military observers in Sector Sarajevo from 1993 to 1994.[3] First, some scene setting is necessary.

THE POLITICAL AND MILITARY SITUATION

Sector Sarajevo was a very active area. UNMOs within the sector counted over 2,000 incoming artillery and mortar impacts and almost 1,000 outgoing mortar rounds on 22 December 1993 alone.[4] Sarajevo could be considered a war zone in all but name at least in terms of volume of fire. However, the political situation was not as clear as it would appear under conditions of a declared war. For those serving with the UNPROFOR there was no identifiable enemy to be fitted in the templates used in North American Treaty Organization (NATO) cold war planning. There were only enemies of the peace, whoever they might be!

In Sarajevo itself, a Croatian brigade had responsibility for holding part of the defensive perimeter against the Bosnian Serbs. Only thirty kilometres away in Kiseljak, Croatians in that pocket were allied with Bosnian Serbs against Bosnian government forces and were even suspected of "lobbing" the odd shell into Sarajevo, hitting on occasion other Croats. On many pre-war ethnic maps, Sarajevo was shown as "colourless" with no definable ethnic boundaries. In 1993, in and around Sarajevo, as elsewhere, all sides were trying to paint the map with their own particular ethnic colour.

Two of the parties, the Bosnian Serbs and the Bosnian Croats, had no international standing as political entities. The issue of whether the fighting was the manifestation of an internal struggle between the legitimate government of Bosnia-Herzegovina (BiH) and rebels from Serb and Croat Bosnian ethnic communities, or rather a manifestation of surrogates acting in the interests of Serbia proper and/or Croatia proper, is still debatable at the time of writing.[5]

As 1993 drew to a close, the military balance, as seen from Sarajevo and the UNMO organization, seemed to have tilted in favour of the Bosnian government in their fight against both sets of rebels or surrogates. The government had the manpower for a larger infantry force than that of both Bosnian Serbs and Bosnian Croats. Indeed, government forces included many Bosnian Croat units/formations.

Grbavica, a Serbian held section of the city core, emerged as a January 1994 flashpoint, which demonstrated the infantry manpower advantage in the urban Sarajevo area. In addition, it illustrated how the government was now taking tactical, and indeed operational, advantage of this military asset. The Bosnian government made a series of night assaults on several widely dispersed locations, throwing the Bosnian Serbs off-balance while drawing the mobile Serb reserves into urban fighting, which favoured fighting on foot. As a result, the Bosnian government was using tactics in operational level actions that nullified the Bosnian Serb advantage in tanks and artillery while exposing the Bosnian Serb lack of infantry.

There were a series of savage attempts to rivet the world's eyes on Sarajevo during this same period so as to force some sort of intervention. These culminated in the market massacre of early February 1994, in which 68 civilians were killed outright while hundreds were wounded (no blame attached). A ceasefire crafted by General Sir Michael Rose, who had just assumed command of the UN military forces in Bosnia, came into effect on 10 February 1994.[6]

General Rose's February peace plan had four main elements, all of which involved military observers. First, there was compliance with the February cease-fire to monitor. Second, there was the verification of the movement of heavy weapons and armoured fighting vehicles (AFVs) either from the 20 kilometre (km) radius Total Exclusion Zone (TEZ) or into UN secured Weapons Collection Points. Thirdly, UNMOs moved into positions between the warring parties until such time as UN troops were available and could do so. In some cases, UN troops subsequently were never permitted in these locations. A major task for the military observers, accepted as they were on both sides of the line, was to determine and report exactly where the line was. Finally, UN observers, again because of better connections with belligerents at local command levels, were asked to facilitate Joint Commission work.[7]

The situation in Gorazde, part of Sector Sarajevo, in April/May 1994 created challenges for the UNMO team leaders involved, as well as for me as their immediate superior.[8] The Gorazde fighting was evidence that the cease-fire was starting to fray. In Sarajevo proper, it seemed that sniper fire was replacing shellfire as a tool for influencing the battle not only physically on the battlefield but in the eyes' of the international media.

This was the environment facing a Sarajevo UNMO, an officer from any one of 39 different nations, from any one of the continents, and more importantly, who may never before have seen snow. As I was told once while on officer training many years ago, nothing is harder to organize than a "bunch" of officers. The officers that I commanded ranged in rank from lieutenant to full colonel, from the army, navy, air force and marines, as well as from branches including combat arms, service support, logistic, communications, intelligence, and no doubt even special forces. Former foes such as Argentine and British marines might serve together. At any one time, there might be only one other Canadian in my UNMO Sarajevo command.

In total, there were between 120 and 200 UNMOs in Sector Sarajevo depending on the date. These were spread among the 12 to 23 OPs/team sites, again depending on the date, which surrounded or were within Sarajevo. The number of UNMOs and team sites peaked just after the February cease-fire took hold. Cease-fires in central Bosnia in mid-March followed by the Gorazde situation led to shrinkage of the UNMO strength in Sarajevo itself. At peak strength there were also 52 interpreters.

UNMOs had to live "on the economy" foraging with their wallets instead of with the barrel of a gun as Napoleon's army had once done. This led, for example, to UNMOs on the Serb side buying beef for the UNMOs inside the city while those inside the city obtained items for those outside from the French post exchange (PX). Some UNMOs were able to arrange for cooking and cleaning. Other teams had to do all chores themselves. Maintaining equipment was a team responsibility, although all technical repairs, in theory, required a visit to the UN field service facilities within

Sarajevo. Technical assistance in location, such as radio repair or vehicle recovery, had to be arranged and often wasn't provided until days later, depending on the situation.

The senior military observer (SMO) reported through the senior UNMO in Bosnia-Herzegovina to the Chief UNMO in Zagreb. UNMO Sarajevo was "in support" of the UNPROFOR Sector Commander, but we were closer to being under operational control (OPCON). A fundamental principle of UNMO operations was that patrols required a mix of nationalities to be valid.[9] Thus, for example, two Canadians would never be together on the same operational patrol. If "politics" is very much a part of the command process at the higher levels of UN operations or indeed any international mission, "people" are at the core of any success of observer operations particularly if the observers are unarmed.

KNOWING YOUR PEOPLE

There is no use "ranting" and "raving" about the quality of the UNMOs that contributing nations have sent. When you assume command of military observers, you must make do with who is assigned.[10] The key for me was following the advice of an experienced former SMO, "Spend your first few days visiting all the team OPs," he said, "Your deputy must have some smarts or he would not have survived. Give him the reins while you get to know your people."[11]

In the case of a Sector Sarajevo, this meant initially, although I could not avoid working at least part time at my headquarters in the PTT (telephone exchange) building where the Sector Commander was located, that I tried to sleep each night at a different team site/OP. I did not manage to visit all immediately. I only managed to visit some of the more isolated sites later in November, as well as making trips to my Gorazde and Zepa teams in that month as well. Certainly these first "sleepovers" brought me great insight as to who was who at each of the key team locations.

It was a practice that I continued throughout my nine months. While others in the PTT adjourned after an evening orders group (O Gp), I often got in my vehicle and drove myself to stay overnight with a team so as to stay current with the people dynamics and to put faces to the voices who answered on the radio net. The frequent rotation of UNMOs dictated such a practice.

Sarajevo, at least in late 1993, was somewhat of an artillery range and a shooting gallery for snipers. Thus, the practice was to post observers to Sarajevo for only three months. This meant that in any one month, at least 30 observers would rotate out being replaced by 30 or more newcomers. Every new UNMO reporting to Sector Sarajevo was interviewed by me personally. "Time spent in recce [reconnaissance] is seldom wasted," is a saying of my first commanding officer (CO). Time spent in getting to know people in an observer mission is likewise seldom wasted.

Personal interviews on arrival also allowed me to lay out some basic ground rules so that newcomers would know exactly where I stood on some key issues. Our business was gathering information, and the words "UNMO Confirmed" meant a great deal in a Sarajevo, where many innocent civilians were killed by belligerents just to obtain CNN sound bytes to wage the international publicity battle.[12] New arrivals were told that a false report meant an immediate posting from UNMO Sarajevo!

Almost of equal importance was security. Shells had frequently landed in the parking lot of the Sarajevo PTT building where my headquarters (HQ) was located.[13] My office window was taped to minimize glass fragments. It had to be clear to each incoming UNMO that violation of my rule of wearing the flak jacket and helmet inside the city as well as on the frontlines and on known sniper targeted routes outside the city, constituted grounds for being sent out of the area. I did send UNMOs from Sector Sarajevo when I observed them in danger zones without their personal protection.[14] Safe driving also had to be emphasized, and new arrivals were asked if they had ever driven on ice or in snow. Simple things had to be checked such as ensuring that the spare tire would be a correct fit.[15] Moreover, while Sarajevo itself could remain snow-free, the surrounding hills, home to the 1984 Olympics, often had deep drifts. Driving up a few kilometres meant a drastic change in road conditions.

I had an open door policy, which I explained to each officer arriving to become one of my UNMOs. Any one of them could walk in but if it involved a complaint against another UNMO or one of my team leaders, then that individual would also be called in so that the issues involved were placed in the open in front of all concerned. Two team leaders were redeployed as result of such discussions while at least one interpreter was moved.[16]

I also had a rule about fraternization with interpreters, which could be grounds for being moved to another team site. Intimate relationships with interpreters disturbed the functioning of the UNMO teams, leading to complaints of favouritism and discrimination. This did not prevent UNMOs from organizing parties and having a good time. Parties were in fact encouraged. Knowing people was a two-way street, and parties offered UNMOs an opportunity to learn about me and about each other in a less formal environment.[17]

One interview demonstrated the value of meeting the incoming UNMOs immediately on arrival. This particular group, from the same Western European country, had the misfortune of seeing the French officer marshalling their UN aircraft after landing at Sarajevo airport, "dropped" by sniper fire before their eyes at the base of the ramp that they were descending. A traumatic welcome to Sarajevo indeed! How each reported his reactions to this event was telling. The one who glossed over the incident in a "macho" manner was the only UNMO that I ever had to move from Sarajevo because the individual could not cope with the stress.

I also had to make clear to new UNMOs that they should not attempt tasks that made them unduly afraid. People not in control of their often well-founded fears are not the best team member for a patrol in a dangerous area. For example, this policy meant that some UNMOs did not do a patrol that inherently involved the UNMO vehicle being shot at every time that their patrol passed a certain landmark.

Courage is the most valued quality in these situations, so UNMOs on a team had to demonstrate this to gain acceptance. Courage is also expected in their leadership. Without being boastful, I must outline some actions taken that seemed to establish my credentials in this regard. There are of course two kinds of courage, physical and moral, and both must be seen by your fellow officers to achieve successful leadership.

LEADERSHIP BY EXAMPLE

I believe leadership by example is critical. Here are some actions that highlight this concept:

<u>Share Danger.</u> During the course of my initial recces and overnight stays at each OP site, I was faced with the same personal risks that the individuals

at that site faced. However, it can mean doing more. When a team leader approached me with a suggestion that a particular team site/OP be evacuated due to the heavy shelling in the vicinity and/or the placement of a belligerent weapon system in close proximity to a UNMO location, I had already experienced life at that site thanks to the overnight stay with that very team. That particular OP had already been evacuated before I took command and re-established once I was in charge. I was reluctant to see this key post abandoned again. Instead of allowing its evacuation, I made it my base for more than a week, living there at night with the team, when most of the firing occurred, and moving from there as required by my duties as SMO. This action had me share many of the dangers, including the exposure moving to and from the OP. I was in a good position to personally assess whether it was too dangerous. The OP was not moved until the Rose cease-fire and decreasing manpower made other sites more suitable.[18]

Be The First to Take Risks. Early in my command, I had to re-establish the right of UNMOs to move across the frontlines at night, provided the belligerents were warned. This of course can be nerve-wracking as belligerent reaction can never be gauged. However, in October 1993, I told both sides, who were engaged in a fire-fight, that I would be coming down the road in two minutes, and then I had to do it. For this first challenge I had to violate my own rules and travel alone. Why expose other UNMOs when it was the symbolism of the UNMO vehicle that was needed? Once this right was established in this and other such forays, I felt easier about asking UNMOs to do the same. Such a right had to be a matter of course, if only to permit casualty evacuation from the Bosnian Serb side to the French UNPROFOR medical facility inside the city. This right of night traverse was to prove invaluable within a month.

Take Risks for Your UNMOs. As the December 1993 darkness fell, one of my UNMOs came on the air with a much shaken voice indicating that he first had been pursued and then subsequently taken hostage by Vasko, a well-known Bosnian Serb terrorist. As he was the team leader at his site, someone else would have to negotiate his release if he was a hostage. I took it upon myself to do so. I had told my UNMOs on assuming command that I would speak up for them immediately when any belligerent party mistreated them. With an extremely competent operations officer (Canadian) on the UNMO net, I launched myself with an interpreter and the Bangladeshi deputy of the UNMOs on the Serb side to demonstrate the value of having already established our right to travel at night. Of course my entire organization was following this possible hostage-taking on our radio net and no doubt all the belligerent parties as well. Fortunately for our well-being, not to mention my anxiety level, time spent on building relationships had paid off and the Bosnian Serb commander in that area had already "gripped" Vasco and secured the release of my team leader before I reached the scene.[19]

In yet another incident involving my UNMOs as hostages, I became a hostage myself. I escaped, not something that I recommend doing, but I then had to return the next day to the same hostage-takers to attempt to negotiate the release of my team, which had been my purpose in going to that location in the first place.

<u>UNMOs in Danger.</u> An immediate response to UNMOs in danger had to be a standard operating procedure (SOP) on the SMO's part. I was practicing this almost to the day I left. Although the ceasefire stopped the shelling in February 1994, sniper fire in fact increased.[20] En route to another site on a late June day in 1994, less than a month before my departure for home, I received a frantic call that three of my UNMOs were trapped in no-man's land under fire from the Bosnian Serb side. Serb retaliatory fire, which had been provoked by the killing of a young Bosnian Serb soccer star, had caught some Bosnian women in their garden plots in no-man's land. My UNMOs had gone to protect them. However, my nearest team commander was not being allowed to see the Serb commander. I sped to this commander's HQ, brushed aside several sentries who pointed their AK-47s/AK-74s at me and was able to secure a stop to the shooting.

<u>Interpreters in Danger.</u> This SOP had to apply to the interpreters who were integral members of the UNMO teams. In one case, one of our interpreters, a Bosnian Serb woman who lived in a village whose men folk were being detained by the Bosnian government forces, was beaten by her neighbours. Besides offering the protection of living in our nearest team compound, we had to make as much a fuss with the Bosnian Serb command hierarchy over this incident as we did over mistreatment of one of our UNMOs.

I was asked why I took such an interest in the fate of one of our so-called "local employees" by the commanders with whom I negotiated. Having owed my own life to interpreters, as they well knew, no answer was really needed. It was important to demonstrate to all your UNMOs that interpreters were part of the team, not simply talking machines or support staff for their personal creature comforts.[21] I had nine months to ensure that all belligerents knew that they could expect me immediately on their doorstep and that of their superiors if any member of the UNMO Sarajevo team, military or civilian, was being shot at, mistreated or being held hostage.

Be the First to Demonstrate the Example Despite Danger. When the TEZ came into effect, there were many reported violations involving allegations of the use of prohibited heavy weapons. An inspection of an explosion in no-man's land was required. I went myself on the first one reported in a particularly tense area. After some trepidation, my Malaysian UNMO partner and I reached the suspected site from the Bosnian side of the frontline. There had been no mines. However some Bosnian Serbs on the other side took shots at us. We had to stand tall, wave the UN flag and carry on; knowing that if we took shelter then the fire would be aimed at us not near us under the pretext that we adopted combat positions. It was important that this standard be demonstrated early in the ceasefire. I had no casualties among my UNMOs in my nine months as SMO, Sector Sarajevo, which I consider my greatest accomplishment.

Moral Courage. Moral courage is also a requirement and for me that meant protecting any UNMO under my command from disciplinary action suggested by higher authorities. Initially, this led to unpleasantness with the Sector Commander who appeared ready to accept the Bosnian government version of an incident on Mount Igman over the report submitted by one of my observer teams. He wanted the UNMO who filed a report that was contradictory to a Bosnian version to be transferred out of Sector Sarajevo. I refused to do this or to change my situation report (SITREP) to conform to what the sector commander said. My deputy, a French speaking officer, found the tension so uncomfortable in the French language only officers' mess that he asked for a transfer. However, the Sector Commander, General Soubirou, came to appreciate both the UNMO organization and me personally.[22]

Similarly, when an observer team made a mistake regarding a Bosnian Serb tank in the TEZ in May 1994, both UNPROFOR headquarters in Bosnia and the Chief Military Observer in Zagreb wanted the offending team posted immediately, as well as their team leader. I said that they would have to post me as well! I was, and still feel, responsible for what UNMO Sarajevo did during my command. Discipline of subordinates must be the prerogative of their immediate superior not reflective of punitive wishes on the part of higher headquarters. Needless to say I did not want to relinquish the most challenging and rewarding job of my military career. However, all the UNMOs had to know who would ultimately accept responsibility. This was not the only time that I had to back up my observers!

My final example: initially, UNMO Sarajevo asked to deploy teams at night to obtain information. I had all such requests passed through me. UNMOs had less special equipment than UNPROFOR troops such as the French and were not in APCs. At night, since origin of fire was often difficult to determine, bored belligerents often shot at UN vehicles. Therefore, the need for the night patrol had to be validated. In most cases, the information gathered, if it could be gathered, would not see any decision-maker until after the Sector Commander's morning O Gp in any case. If there was a valid reason, then usually I, or my Deputy, or one of the team leaders would undertake the task. This I made clear to whoever made the request, which was usually funnelled through sector operations rather than my own headquarters.

Moral courage was also required to support the interpreters, classed at that time as local employees on month-to-month contracts. I literally ambushed a VIP from UN headquarters New York as the individual emerged from a meeting to plead for the UN to provide medical support for a Bosnian Serb interpreter from my Sarajevo team. This interpreter did not trust Croatian facilities in Zagreb, was not acceptable to hospitals inside Sarajevo, and could not receive the appropriate treatment from embargo deprived Belgrade facilities. In the meantime, Bosnians from inside the city, with no apparent UN connection, were receiving treatment at an Italian hospital in Ancona, Italy at the other end of the international air bridge.

Courage was also related to earning the respect of the belligerents as development of a good relationship with commanders of the various factions played a role, at least in 1993/94 Sarajevo, to UNMO survival.[23] Proceeding with a Bosnian government commander to the scene of a reported incident in my vehicle, I was warned by my nearest OP that the road ahead was being shelled. The team recommended that I turn back. I asked my Bosnian liaison officer if he and his commander wished to turn back. "No," they said. As a result, I had to proceed!

Early in the ceasefire, the Bosnian Serbs reported a violation in the form of Bosnian government patrol in a ruined house. It was important to act quickly on such complaints so with the Bosnian Serb liaison officer I proceeded to the edge of no man's land, at which point he refused to go further because of the threat of mines and unexploded ordnance (UXO). I went ahead with my interpreter because it was important to verify this allegation. I simply relied on my experience in mined areas to warn me of danger. I was able to verify that there was no Bosnian presence. This physical demonstration earned respect, which in turn facilitated my relationship and that of my teams with the belligerents.

Moral courage was also required when dealing with the belligerents. Once again, early in the ceasefire I received a large influx of UNMOs from Muslim countries. One Bosnian Serb brigade initially refused to accept a Jordanian officer. I did not accept this, and I am happy to report that this same brigade requested that this UNMO from Jordan be extended beyond his six-month tour when the time came to move him.[24]

RECOGNITION

Getting to know people was important so as to understand the person who spoke on the radio or landline. It was also important so that the qualities needed for team leadership and key staff positions were quickly recognized and the right individuals were selected before they were posted out. For example, proficient English speakers were needed on the UNMO Sarajevo net. Time lost in "saying words twice" might mean an important message, such as details of ongoing shelling or sniper fire, or a casualty, was jammed out. I paired a Bangladesh MIG pilot who was great at picking up key items from English radio traffic with a Bangladesh Army Intelligence officer less used to oral English, particularly on the radio, to form an operational team. It worked.[25]

In an UNMO organization, I felt that it was important to discard any prejudices that I felt about the skill-sets of any nationality. Everyone had something to contribute. The key was to find what their contribution could be. Some with almost no English and no driving skill, particularly on mountain winter roads, could only demonstrate their value when shelling and sniper fire was intense, displaying a level of courage that inspired colleagues with a wider range of skill sets.[26]

Recognition of the contribution of UNMOs was important. I made it a point to host my immediate staff and subordinate team leaders with their colleagues at a farewell dinner when they were leaving. At that time, I gave them a plaque made locally that I designed and paid for with my own funds.[27] In addition, all UNMO evaluation reports required SMO comment and the best recognition that an observer commander could give was to dedicate time to write what each UNMO deserved on this document, which was often used to determine future postings especially for those posted for a year, or longer, as an UNMO. Some of the assessments, particularly those of team leaders and on personal staff, required completion almost entirely by the SMO. I made it a point to show these evaluation reports to the individuals prior to their departure so I could do a final interview. This deadline served to motivate team leaders to do their part in this evaluation process promptly. Since I did these on my own computer, I kept copies of my comments, which proved invaluable as I was asked several times after returning to Canada to provide some input on the performance of particular UNMOs who served me in Sector Sarajevo.

I also learned to recognize what was important to my teams. For example, UNPROFOR Headquarters gave Benazir Bhutto was given the "red carpet" treatment by on a visit to Sarajevo. My contingents of Pakistani UNMOs were upset when Nawaz Sharif, potentially the next President of Pakistan, was not accorded the same orientation to operations in Sarajevo.[28] With the resources of UNMO Sarajevo I tried to remedy that deficiency while allowing Pakistani officers face time with Ms. Bhutto's main opposition equivalent to the time that they had been given to participate in her visit.

CONTACT

Physical visits had to be continued and frankly these were no hardship. In my headquarters, I slept in my office and ate at the French Army cafeteria. I ate and slept better at the team sites while learning about my UNMOs. I also used the radio. Information was a two-way street and my UNMOs had to know the situation immediately. Details of shelling (impact areas – not necessarily who was responsible even if known) sniper fire (locations under fire not who was shooting if known) and mines were passed immediately. After every negotiation that I was involved in, I briefed my organization via the radio immediately after emerging from the meeting. This had the effect of not only keeping my entire organization in the picture but also of passing on to the belligerents what I thought had been said at the meeting as all parties to the conflict monitored all nets.[29]

It may seem difficult with UNMOs scattered over the mountainous terrain but mass briefings when possible were invaluable. For example, when the February 1994 ceasefire suddenly "kicked in", I could not personally interview all the 50 or so new arrivals sent for a short-term augmentation. Before I dispatched them to their various teams, however, and, over the protests of the French staff that resisted relinquishing the only large room in the headquarters, I gathered them all together for briefing.

Initially, the task assigned to these newcomers, as well as to my already deployed UNMOs, involved around the clock surveillance, or rather listening for artillery violations. I followed up my initial mass briefing with midnight and early morning (0200 hours) visits to the team sites where these new arrivals were "listening and looking" in the dark. The only darkness should be physical not "situational" as the senior UNMO must ensure that everyone knows as much as he does. On these visits the information exchanges benefited us both while permitting contact with individuals whom I had not interviewed.

PERSONAL SUPPORT

If you know your people you are not alone. My Malaysian deputy, a parachute veteran of counter-insurgency operations in the jungle, shared the task of visiting during the night, and indeed a whole range of dangers, while displaying leadership qualities that left me confident that I could enjoy my leave time with UNMO Sarajevo in good hands. A British marine, followed by a Canadian infantry officer, a Dutch air force officer and then a New Zealand pilot as my chief operations officer ensured that I was well supported by my headquarters operationally. A Dane and a Ghanaian, among others, made sure that our logistics was not a cause for complaint. Two Bangladeshi officers ran my personnel system. All had to have my unqualified support. When required, I would take up one of their issues to a higher headquarters.

Again I was very well-served by most of my team leaders.[30] It was important to remember that most UNMOs are officers with leadership qualities. Thus, I followed the rule of maximum delegation in local matters supported by personal intervention on my part when required. For example, when the ceasefire took effect, the Kenyan officer leading one of my teams managed to create freedom of movement for the UN through a hitherto blockaded/mined road into Sarajevo. His initiative led to more patrolling once the previous much longer approach to this area was no longer required.

In another case, when the NATO air-strikes at Gorazde stopped contact with the Bosnian Serbs, a Jordanian team leader used this lull in normal activities to initiate a garbage cleanup in his area of operations, something of course, well outside our UNMO mandate but done primarily by locals with his leadership. Rarely was I disappointed when I delegated!

Support from above was also outstanding. Colonel Pastoor, the Dutch Colonel who was my UNMO superior in Bosnia, had served with me on an OP on the Golan in an earlier UNSTO tour. He was consistently supportive. The Chief UN military observer (CMO) in Zagreb, Brigadier General Pellnas, a Swede, had been my first CMO in Afghanistan. UNMO Headquarters in Zagreb, particularly the UNMO Personnel Officer who backed all my posting requests/rejections, provided consistent and timely support although at times he had over 1,000 UNMOs for whom he was responsible. Colonel Pastoor was followed by Colonel Tairobi from Malaya, who continued to give me the same level of support rendered by his Dutch predecessor. General Sir Michael Rose, his personal staff, and General Soubirou with his personal staff, gave me tremendous support as well.

PERSONAL COMPETENCY

Liking to deal with people is the first requirement. It is hard to imagine a Canadian officer without enthusiasm for that task. To command UNMOs, as mentioned earlier, a leader must know as much as possible about each individual observer and all the belligerent commanders who have any bearing on their security. Developing an understanding of these people will be a challenge. At all times, the SMO must guard against personal prejudices becoming a smoke screen obscuring the real qualities of the persons involved, either the officers that work for him or those with whom he, or she, must negotiate.

My own Sarajevo experience, complemented by Afghan and South Lebanon observer time, has led me to believe that every observer has something to contribute. Moreover almost everyone, including belligerents, have something about them to admire. The trick for the SMO is to discover these![31]

Competency also involves knowing the ground where your UNMOs will work under conditions of both day and night. Today, this is made easier by the global positioning system (GPS) but you must still be able to navigate on the "used" routes in total darkness, without lights if that is how your observer must do it. There is little place for the UNMO leader who is weak with a map – even if they have the most proficient map-reader as a commander's driver. There is no excuse for not knowing the terrain where UNMOs must undertake their mission, as well as all the belligerent positions from where they can expect danger. On the other hand, any UNMO who claims that he or she was never disorientated, or lost, cannot be trusted!

MISTAKEN JUDGEMENTS.

Some of the disappointments of my command in Sarajevo relate to my inability to maintain close contact with all my team leaders and those that worked for them. The Gorazde and Zepa teams, located hours away in Eastern Bosnia, were difficult to visit as checkpoints resulted in delays, often for hours and in some cases for days.[32] Time spent in either pocket involved time away from Sarajevo. The hostility of the Bosnian Serbs at the famed Rogatica checkpoint where delays were so common relates directly to the difficulty in maintaining contact with these two teams.

Unknown to me before his last few days in Sector Sarajevo, an UNMO from the Gorazde team, not from either Asia or Africa, had been involved in smuggling money into the pocket. Not only had his action resulted in stricter searches, often in violation of the freedom of movement that the UN was supposed to enjoy, but his conduct threatened the status of all the UNMOs on the Bosnian Serb side of the line once known, as it soon was. I relieved the Gorazde team leader of his responsibilities, who had been there when I arrived, and had him posted out of the Sector. In two, all too short visits to Gorazde prior to this, I had not picked up on this activity.

One of my Zepa team leaders, an outstanding military officer, asked to be relieved as he did not want the burden of handling the responsibilities of the UN High Commissioner for Refugees (UNHCR) representative's duties in the frequent absences of that individual.[33] Again, I had not visited Zepa during this officer's appointment as team leader there so it was a disappointment to me as well as a problem in selecting a replacement with the qualities that suited.

My open door policy, however, helped ameliorate the sheer difficulty of maintaining contact. When one of my more isolated Sarajevo teams was upset by dissention, one member was able to initiate a face to face meeting with the team leader, and himself, in my office, as result of feeling secure enough to avail himself of this policy that had been made clear to him on his arrival interview. This particular team's situation and that of other teams with which I had missed something in my contacts/visits were sometimes remedied as a result of such meetings.

Some situations were missed altogether. Often these involved personal relationships that impacted their teams. One UNMO had developed a relationship with an interpreter against my policy of non-fraternization, which I only found out about later. Another similar but more public case occurred in Zepa. A New Zealand UNMO, no longer in the Sector, or under my command, managed to smuggle out his Bosnian girlfriend from the Zepa pocket with the help of some UNMO colleagues. I did not know the extent of the complicity among my own UNMOs until I read his book, another disappointment.[34] The immediate consequence was that an UNMO, from a NATO country, who had helped him was no longer acceptable to the Bosnian Serbs as a liaison officer, an appointment for which I had selected this particular officer, a job for which his background suited him. That was how I initially discovered the incident, although I was not aware of all the details that later appeared in print involving criminal activity.

ON THE BORDERS OF RIGHT AND WRONG.

Sometimes there are things that you do that fall in an ethical grey area. One such incident involved knowledge of an extensive tunnel system on the frontlines. For months, my personal reconnaissance on both sides of the line had led me to believe that the Bosnians had constructed a substantial tunnel through which an assault could be mounted on the thinly held Bosnian Serb lines opposite their own. In almost my last days in Sarajevo,

I was shown this tunnel system by the local commander who knew that I was leaving on my word that I would not divulge the exact location. Curiosity led me to agree. Subsequently I did indicate to my successor that that was an area requiring careful scrutiny due to the presence of such a tunnel system without giving its precise grid. There were other such incidents involving team members. The commander must be able to live with himself and do what he, or she, thinks is right. There are few who can offer practical advice to the SMO on the matter of ethics.

LETTING GO!

There is no escape from the fact that once you know people, and they know you, that it leads to development of a close relationship, not necessarily positive in all cases. Few Canadians will have shared your experience, so return to Canada from commanding UNMOs or observers can be "downer". The situation on the ground may unfold for better or worse, but you will find yourself wondering about how life has impacted on those whom with you worked so closely with such an investment of your own person.[35]

PREVIOUS EXPERIENCE OR THE INTERVENTION OF FATE.

Several prior posting contributed, in my opinion, to my ability to serve nine months in Sarajevo as Senior UN Military Observer, the longest of any SMO, without a mental breakdown, to the satisfaction of my superiors, and with no casualties among my UNMOs.

I had served in Cyprus as a Reconnaissance Troop leader in a Canadian independent Reconnaissance Squadron that had been trained up for this task by Colonel D.A Nicholson of "Where Have All the Tigers Gone" fame. I had prior UNMO tours on the static OPs of the Golan, served in Jerusalem as UNTSO Operations Officer and indeed had been hi-jacked in South Lebanon. A tour in Kabul as an observer in 1989 during a period of many rocket attacks prepared me for Sarajevo shelling. I also commanded a tracked armoured reconnaissance squadron in Canada, earmarked for Canadian Air Sea Transportable (CAST) Brigade duty, served as operations officer for the Canadian Forces Europe (CFE) Fallex Umpire organization, and ran the operations centre for the Rendevous 1983 umpire unit that no doubt contributed to my suitability as an SMO in both overseeing collection and processing of information.

Furthermore, just prior to assuming command in Sarajevo, I had been leader of a team of 25 UNMOs in Macedonia based at Orhid. I had some struggles turning individuals into a team rather than permitting them to treat this as a relaxation and recreation (R&R) break from other more onerous assignments in UNPROFOR. One practice that I had instituted there was a daily briefing on the Sarajevo situation, preferably by someone who had served there, which served me well later.

An oft-overlooked preparation for UN, indeed international service, is a tour at the National Defence Headquarters (NDHQ), which familiarizes you with the political environment found in a politically sensitive military headquarters, which indeed all international headquarters tend to be. In the end, it was a matter of chance that Canada provided me with the opportunity to be an UNMO in the Former Yugoslavia, let alone serve as SMO.[36]

SELECTION

The selection of observers by the Canadian Forces, indeed the selection of many individual staff officers for individual positions, including some of the highest is too often left to chance and the career manager, perhaps with the input of the senior serving officer of the appropriate branch with his personal agenda.[37] There is often no objective assessment of personality, skill sets or experience, or indeed fitness level that might be called for by the observer assignment. Other countries note this. During the so-called "Dark Decade" of the 1990s, Canada's image was saved on observer missions and individual staff appointments by the outstanding initial selection and training provided our junior officers and their subsequent staff experiences.[38]

SOME CONCLUSIONS

In closing, I offer these words of advice:

<u>Dedicate Time to Knowing Your Observers.</u> Military observers and the information that they provide is the weapon system of the observer organization. Eight years after I left Sarajevo, I was testifying on the basis of UNMO Sarajevo reports at the trial of a general indicted for war crimes in the conduct of warfare around that besieged city. No soldier would neglect his or her weapon. Similarly no commander of observers should

neglect the observer personnel under his or her leadership. Getting to know your people starts upon their arrival and is an ongoing process that does not even end when you recognize their performances with an evaluation report and more. Knowing people means that contact will continue years later!

People Must Know You. No one will risk their life for a cypher. You must create opportunities for your teams to know you. Part of knowing you must be the example you provide. Everyone will have their own style. Visits, an open door policy, and often being first on the scene of a possible hot spot helped me.

Dangers. Leadership is essential in facing danger. On almost all missions, military observers are unarmed. The senior observer must be the first to take risks and to demonstrate dangerous survival techniques. Danger must be shared and risks taken to protect your observers, and interpreters, that are in danger. In this context, remember the tendency of headquarters to be in somewhat more safe locations than isolated OPs/team sites. Orders that ask your teams to undertake unnecessary risks must be challenged. This exposes you to dangers of a different kind. If the job must be done, do it yourself. That usually makes the requester pause! There may be other non-physical dangers in standing up for your teams, observers and interpreters.

Stand up for Your Observers (and interpreters). This principle, no doubt is still taught to Canadian officers in basic training andmust be followed by any commander of observers. Any attack on one of your observers must be considered an attack on you, whether in the field or in the headquarters. In the field, you are usually dealing with belligerents. Your personal relationship with their commanders is part of your protection plan for your subordinates.

It is not so when the finger of blame is being pointed. Discipline is the responsibility of those who command in the field. In any observer organization, the proverbial "buck" must stop at your desk, not descend in torrents on those below. The flip side is that this also means that the observer commander must act when one of his or her observers violates rules clearly enunciated on arrival on the mission. However, I should add that of more than a thousand UNMOs who served under me in my nine month command in Sarajevo only a handful had to leave by my direction. No one can enter into personal relationships forged under somewhat challenging circumstances and remain untouched and more to the point unchanged. People performing for the good of "UNMO Sarajevo" made the command of observers there the most rewarding tour of my thirty-five years of CF service.

ENDNOTES

1   UNMO is used for UN Military Observer throughout the remainder of this chapter. However, the term now favoured is "military observer" or "milobs", in recognition that so many observer missions are not UN sponsored. In August 1993, the Bosnian Serbs pinched off the Bosnian supply line which crossed the heavily forested Mount Igman eventually reaching besieged Sarajevo through the famed airport tunnel. A UN ceasefire had brought about a withdrawal of the Bosnian Serbs once again permitting the flow of supplies albeit restrictedly. Incidents continued as the Bosnian Serbs harassed the movement of people and items while the UN attempted to find the best positions from which to monitor what was happening.

2   The French Sector Commander, a former Legionnaire, was particularly upset because at this point the Bosnian Army or the mutinous segment represented by the hijackers were refusing to release the French APC although all UNPROFOR personnel had been let go, but without their personnel kit or possessions.

3   General Sir Michael Rose, *Fighting for Peace*, (London, The Harvill Press, 1998), 57 was very complimentary of my performance in his book so I must have been doing something right.

4   This detail and the chronology of events that the author gave the International Criminal Tribunal for the Former Yugoslavia (ICTY) as part of his testimony in November 1998 to an ICTY investigator is available on-line by referring to the author's appearance in the Galic trial, IT-98-29, 30,31 May and 3 June, 2002 and the Milosevic trial, IT-02-54, 13 and 14 November, 2003, at www.un.org/icty

5   At the time that the author appeared at the trial of Milosevic IT-02-54, 13 and 14 November, 2003 the ICTY Prosecution was attempting to draw a connection at that time between the former Serb leader and the shelling of Sarajevo.

6   Roy Thomas, "Implementing the February 1994 Peace Plan for Sarajevo" *Canadian Defence Quarterly*, Volume 24, No. 3, March 1995, 22-26, outlines details on this ceasefire which held until the Summer of 1994. General Rose took advantage of the first Sarajevo

CHAPTER 1   IN HARM'S WAY

market massacre on 5 February 1994 to utilize public relations to fashion a ceasefire. The second Sarajevo market massacre was the trigger for the NATO air campaign, Operation Deliberate Force outlined in Tim Ripley's *Operation Deliberate Force* (Lancaster: Centre for Defence and International Studies, 1999) Chapter 19, 228-244.

7   Roy Thomas, "Sarajevo UNMOs" *esprit de Corps*, Vol 4, No. 1, 1994, 9 outlines specifically the role UNMOs played in Sarajevo in implementing Rose's peace plan.

8   Roy Thomas, *From Gorazde to Kosovo: Aerospace Power in Peace Support Operations*, (Toronto: Canadian Institute of Strategic Studies, 2000).

9   While this is true on UNMO missions, other missions such as the short-lived Kosovo Verification Mission (KVM) 1998/99 had members of the same nationality working together although civilian and military personnel were mixed.

10   From personal anecdotal evidence on UNMO service on the Golan, Lebanon, Afghanistan, Macedonia and Sarajevo, will attest that so-called Western observers are no more honest or courageous than observers from other countries.

11   Norwegian SMO, Macedonia, September 1993.

12   In June 1994, at 1400 hours, an old man was shot by a sniper, who could have been either Bosnian Government or Bosnian Serb. The only motive appeared to be to have this death appear on CNN. While the body lay there a media team passed by, and stopped, with the result that this killing attributed by the media, to the Serbs, was a news item in North America during the 1800 hours prime time evening news. UNMO reports provided UN authorities including the Secretary General himself, the facts or lack of facts relating to this allegation and so many more. The UNMO nightly SITREP reflected what UNMOs reported particularly in the case of casualties, and did not necessarily agree with the Sector Sarajevo SITREPs, which on occasion was a sore point. However at my level and at the team level, reports had to say what we saw and say what was not seen. Conjecture had to be labeled as such. New arrivals were told that a false report meant immediate removal from my command. During my tour I did not have to relieve any UNMOs for this reason.

13   From my PTT window, I personally saw a mortar round explode 10 meters from two UNMOs just as they were getting out of their vehicle.

14   Some of the insurance coverage for Norwegian observers was not valid if they were injured or killed while not wearing their helmet and flak jacket which was an issue when General Rose asked UNPROFOR personnel, as a confidence measure, to stop wearing protective clothing all the time outside.

15   Roy Thomas, "AAA, Safety-UN Military Observers (UNMOs) in Sarajevo," *DND General Safety Digest*, Edition 1, 1995, 12-14.

16   Roy Thomas, "U.N. Military Observer Interpreters in Sarajevo" *Language International*, Vol. 7, No. 1, 1995, 8-13.

17   For example I hosted, with some help from Colonel Wlasicuk's Strathcona kitchen, Canada Day activities for my UNMOs both inside and outside Sarajevo to include locals that they were involved with.

18   At "Papa Three," they had an excellent cook, the landlord himself, who was excused military duty to keep an eye on us and his property. His son collected the signatures of every UNMO who stayed overnight at this site.

19   Roy Thomas, "The Role of Relationships in Providing Security for Unarmed UN Military Observers" *The Liaison*, Vol. 2, No. 3, 2002, 28-33.

20   Thomas, "Implementing the February 1994 Peace Plan," 24.

21   Roy Thomas, "There's No I in Team" *The Liaison*, Vol. 3, No. 1, 2003, 36-44.

22   Then Major, now Brigadier-General, Dave Fraser, executive assistant to General Soubirou told me at the time that the Sector Commander had cancelled engagements to host a private farewell dinner for me, a far cry from the coldness I had encountered when I first challenged this former Legion officer.

23   Thomas, "There is no I in Team," 29.

24   On a personal note, I no longer drink and from the first meeting I had to establish that I did not want alcohol under any circumstances. For those who feel that having a drink was necessary in negotiations with Bosnian Serbs I can only say that not only did I manage nine months without taking a drink but was treated to a dinner with the Commander and officers of one of the Bosnian Serb brigades, without alcohol being served as a farewell gesture.

25   UNMO Headquarters in Zagreb stole my MIG pilot for UNMO operations there. In fact, many of my operations duty officers moved to other HQ. It should be noted that selection for team leading and UNMO headquarters staff positions only came after tours, some shorter than others, with teams deployed around the city.

26   I dismissed a team leader, with previous UNMO experience on another mission for

CHAPTER 1

treating one of his team with discrimination/prejudice. Another very experienced UNMO called the Pakistani UNMOs "ragheads" which not only demonstrated his prejudice but also his ignorance of that area of the world. There were no officers from India or Sikhs on the mission! As a graduate of the Pakistan Army Command and Staff College I was often appalled at how little Western officers knew about recent combat experiences of colleagues from other continents.

27   I gave departing team leaders and key staff members copies of Major-General Lewis Mackenzie's book, *The Road to Sarajevo*, which had recently appeared on bookstore shelves. One UNMO, then Captain, now Major Don Haisell, collected signatures of all his comrades as well as the autograph of Mackenzie himself and General Sir Michael Rose.

28   Christina Lamb, *Waiting for Allah* (London: Hamish Hamilton, 1991) outlines the political situation in early 1990s Pakistan including the place of Ms. Bhutto and Nawaz Sharif . It also throws light on events in Afghanistan at that time, which have implications as this is written.

29   I was told by my Serb interpreters that I had been on Bosnian Serb TV but what had occurred was that my radio traffic had been recorded, edited and then dubbed in with completely different film footage. "Psyops" was a factor in 1993/94 Sarajevo.

30   Thomas, ICTY testimony. A list of key personnel that I dealt with is found in my November 1998 witness statement subsequently entered as evidence in both trials.

31   Younger officers from armies with our tradition will not have experienced NCOs to help their judgement. I can only say that it is hard to like or be civil to someone you know wants to kill you!

32   The Ukrainian Deputy Commander of Sector Sarajevo, who ironically had served in Kabul the same time-frame as I had in 1989, was detained in February, 1994 at the Rogatica checkpoint for almost a day in his jeep.

33   The UNHCR representative in Zcpa was absent from his post in that pocket for six of the nine months of my tenure as Commander of UNMO Sarajevo. This meant that either my UNMO team leader or the UNPROFOR company commander acted in his absence. My UNMO team leaders starting with a Dutch parachute veteran, became very involved, so much so that the 8000 refugees who owed support to him called their camp, UNMO Village.

34   Sue MacCauley, *Escape from Bosnia* (Christchurch: Shoal Press, 1996) is a compilation of the tangled plot from the perspective of the two so-called lovers. They are not together at time of writing.

35   Meeting other former SMOs of Sector Sarajevo at the ICTY in den Hague as well as being part of the witness process helped bring closure. I am in contact with many of my comrades from Sarajevo.

36   An army colonel tried to prevent the author's posting to UNPROFOR apparently on the basis of having had too many tours already and because of my age. Fitness should be the factor, not age. In 1992, I had just completed several triathlons and in 1993, was training for further participation in such events. Other observers have commented to me personally on both immature and out-of-shape Canadian officers selected as UNMOs.

37   The author has already noted in another CFLI book on leadership, how the selection of Colonel George Oehring as Commander of Sector South in the Krajina, where his performance was recognized by a Meritorious Service Cross, was purely an accident.

38   During my tenure at CFB Suffield as the Range Control Officer, I personally observed 29 Medicine Man serials. At the platoon level, Canadian platoon commanders were better trained than their British counterparts. The excellent branch training given combat arms officers accounts for the excellent performance of so many Canadian observers almost in defiance of the lack of selection policy for observers prevalent in the early 1990s. I cannot speak for the performance of Canadian observers today, so many of whom have not had the benefit of what used to be, and may still be, almost the best combat arms officer candidate training in NATO.

CHAPTER 2

OPERATION TOUCAN OVERVIEW*

Rear-Admiral Roger Girouard

Operation (OP) TOUCAN was Canada's joint force contribution to the United Nation's (UN)-sanctioned, Australian-led, coalition that landed in East Timor on 20 September 1999, to quell the violence that had erupted in Indonesia's 25th province in the aftermath of a pro-independence vote. The impetus for the mission arose from a number of sources that included the Prime Minister's presence in Asia for trade talks, the UN's requests for support of the vote that it had assessed as free and fair, as well as the Australian Government's strong requests for support and assistance in its plans to stabilize the situation in East Timor.

Given the fact that the Canadian Forces (CF) were already heavily committed to international peacekeeping and to the domestic imperative of OP ABACUS,[1] a number of compromises in developing the force mix were made, which in the end, provided for one of the most unique and eclectic missions in recent memory. In totality, the three services and many support elements, in all some 750 Canadian service personnel, were located in six sites, stood up, achieved a steady-state of operation, served two administrations, namely the International Force East Timor (INTERFET) and United Nations Transitional Administration East Timor (UNTAET), and effected redeployment without relief, all in the span of six months.

BACKGROUND

Delving deeply into the history and ethnic complexity of Indonesia is an area of study all unto itself, but that is not the aim here. Suffice to say that Indonesia is a large and populous island nation with many diverse ethnic groups. It has been subject to both a colonial past as well as the Muslim expansion into southeast Asia. East Timor is the remnant of Dutch and

---

* A version of this essay was originally published in the *Canadian Forces Maritime Warfare Bulletin*.
[1] OP ABACUS was the CF planning for (and potential reaction) to the new millennium computer scare.

Portuguese colonies and it was the site of a small guerilla war against the Japanese in concert with Australian forces during World War II. Later, upon the relatively rapid departure of the Portuguese in the late 1960s, it was invaded by Indonesia.

In the subsequent 30 years, East Timor remained the poorest region in Indonesia. Its population was largely Christian; rural farmers and labourers. Technical, management and administrative skills and positions were generally retained by the Jakartan Indonesians, who controlled business, investment, government and schools. Many religious schools operated, providing basic skills and even some schooling in English, but these could also limit the potential of the ethnic East Timorese. Some parallels to Quebec before the "Quiet Revolution" can certainly be made.

East Timor's transition was, however, less than quiet. The UN-monitored vote for independence on 30 August generated an immediate and significant response from the Indonesian Army (TNI) and local militias. Indeed, pre-vote violence and intimidation were such that veteran UN staffers were taken by surprise and feared for their own safety during the ballot. With the announcement of an overwhelming choice for independence, the rampage, as it came to be called, commenced in earnest. With its trigger in Dili, which was the province's capital and largest urban centre, anti-independence violence spread first to the other large centres and then on to the countryside, following highways, roads and country lanes.

The planning effort, energy and logistic skill of this campaign of destruction would come to amaze everyone who would arrive in East Timor in the coming months. In some centres, as much as 80 percent of the infrastructure was destroyed. In a third-world country where many people had only a grass hut, a pig and a pot, even these were taken away, like some bizarre imitation of the "Grinch who stole Christmas." So systematic and complete was the effort that fire trucks were filled with fuel instead of water and driven down back roads and cattle tracks, so that they could better ignite dwellings in village after village. Reports of TNI participation, let alone aiding and abetting the militias, were common. Livestock was killed or collected for transport to West Timor. Villagers were driven into the hills or forced into trucks heading west to internment camps. At times marauders gave a few hours warning to allow people to pack up some goods. Often, villagers were simply forced to run for their lives, splitting families in the panic. Those who could not flee were left behind, sometimes never to be seen again.

Given the UN presence and a significant media interest, the eruption of violence was quickly placed on the world stage. Australians clearly had an interest, for economic and security reasons, in the events that were literally taking place in their backyard. Nonetheless, an issue of conscience was also at play given the support of the Timorese against the Japanese in the Second World War, as well as a sense at least in some circles in Australia, that East Timor's occupation by Indonesia had not been Australia's finest hour. Such was the belief that something had to be done to quell the rampage that plans were developed with an eye to Australia going it alone in early September. However, world opinion, UN and American diplomacy, regional responsibilities and bilateral and Commonwealth markers all contributed to the formation of an Australian-led coalition of some 20 countries. Its insertion into the city of Dili by sea and air on 20 September marked an international response to a military problem, as opposed to a reaction based on a purely humanitarian crisis. Nonetheless, the reaction was instantaneous by any standard in recent history. The fact that this rapid response was also comparatively robust, the many challenges of logistics notwithstanding, is also noteworthy, and a point that was not lost on the TNI or the Government of Indonesia from the outset.

THE CANADIAN ASSESSMENT AND COMMITMENT

Many factors played into Canada's participation in what came to be called the International Force East Timor, including the Prime Minister's presence in Asia for trade talks, the realities of pressures from the halls of the UN, as well as the significant lobbying from both the Australian and New Zealand Governments. The issue in early September quickly became less of "would we" than "with what."

The army's commitments internationally and to OP ABACUS precluded a battalion or battle-group from being tasked, so the focus quickly and appropriately shifted to air and maritime capabilities. The two C-130 Hercules transport aircraft that were assigned were a clear requirement for the coalition and relatively easy to provide. Her Majesty's Canadian Ship (HMCS) *Protecteur* too, given the successes of disaster relief missions by Auxiliary Oil Replenishment (AOR) ships in Somalia and the Bahamas, was a viable and flexible platform for the mission as envisaged. It should be made clear at this point that the mission plan was developing by means of conversations with the CF Attaché in Canberra, who was translating early assessments of potential roles and tasks from the Australians. Detailed plans, and a strategic reconnaissance by Canadian staffers, were still to come. The vision for the employment of HMCS *Protecteur* was for her to be relatively Dili-centric. The large ship was intended to provide a medical team ashore, humanitarian aid and to serve as a floating command post for the Canadian Joint Task Force (CJTF) Commander. This latter aspect helped contribute to my selection as commander of the mission.

Defining the land element itself was the true challenge in forming the force mix for OP TOUCAN. Other national commitments limited available numbers of troops and sustainment into the long term was seen as problematic, so the concept of a company group was developed. Initially, the aim was to deploy the force in Dili to provide security to Canadian elements operating there, including the medical teams and the humanitarian effort ashore. I envisaged the plan to be more akin to fixed point security such as the camp at CANADA DRY in Qatar during the first Gulf War as opposed to the security operation conducted in Haiti.

Clearly, the Australians were seeking as many "bayonets" as possible given the potential threat levels in East Timor, the residual TNI and militia presence, and the distinct possibility of a protracted guerilla war. The Canadian army's experience, robustness and similarities, in view of common roots in the British army system, meant any land contribution would be appreciated, and likely placed in the thick of it. In time, this would play on the actual assignment and location of the as yet to be defined company group.

Those who have experienced the sudden shock of the starting gun for a major deployment, such as we saw for the Gulf in 1990 for instance, will appreciate the organized mayhem of technical preps, staff planning, deployment screenings and mission briefs that fill the scant hours between the warning order and letting the first elements go. Although of a different scale than the pre-Gulf preparations, the efforts by the fleet maintenance ship *Cape Breton* on HMCS *Protecteur*'s behalf are no less noteworthy in my view. A particular focus was given to the control, communications, computers and intelligence (C3I) suite in view of the location of the JTF staff on board. Maritime Pacific Headquarters' (MARPAC) Naval Constructor Troop was also on alert for any of a number of functions which might come their way.

At the time, my own role was simply serving as acting fleet commander, working to help HMCS *Protecteur* get out of harbour fully equipped. One day, after the morning brief, Commander Randy Maze and I were discussing the progress of preparations with Vice-Admiral Ron Buck in his office when he turned to me to ask if I knew who the mission commander would be. I think I reflected the "deer in the headlights" stare as he advised me that my name had been put forth. A tremendous compliment and an easy job to say yes to, although the role we had in our minds at that point had little resemblance to what eventually played out.

It's fair to say that your level of attention goes up when your place in the task at hand is clarified as it was for me. It would be wrong however to suggest that one's ability to influence events or decisions in the larger scheme of things is enhanced to any significant degree. We are accustomed in the navy, as ship's captain or even task group commander, to having some measurable impact on the planning process and mission development. Now as a JTF Commander-selectee in the forum of a national joint operation, the reality is that my horizon of interest exploded beyond MARPAC now to include an Airlift Task Force (ALTF) and the still somewhat nebulous entity that was being formed by the army. Though volumes of information were now flowing my way, the lead in the effort to develop OP TOUCAN clearly and appropriately resided with the former offices of the Deputy Chief of the Defence Staff (DCDS), specifically Joint (J)3 Operations (Ops) and the J3 International cell, with input evident from both the Department of National Defence (DND) Policy world and Foreign Affairs Canada (FAC).

At this point in the mission development, a Strategic Recce Team was dispatched to Australia to undertake more detailed discussions with the Australian Defence Force (ADF). It was led by a combat arms lieutenant-colonel; he was accompanied by Lieutenant-Commander Marc Fletcher, who was assigned to provide Maritime input. On 20 September, while this team was swinging through the Australian national and army headquarters in Canberra, the landing by INTERFET actually took place. Concurrently, my J2 designate, Lieutenant(Navy) Owen Parkhouse from Maritime Atlantic headquarters (MARLANT) and I were making our way through the operational cells of National Defence Headquarter's (NDHQ) J3 and J2 (intelligence) elements to be briefed and brought up to speed on the discussions with Australia and their impact on potential force composition. Ironically of course, the reality from the coal face in Australia was not always fitting into the concepts which we had already begun to feel comfortable with here in Canada, but it was still early days and much would change yet. Meanwhile, the realities of Task Orders, rules of engagement (ROE), staffing process, finance, Table of Organization and Equipment (TO&E), chain of command, transfers of operational capability assignments and a mountain of other subjects were all being covered and discussed at seemingly lightening speed.

Of course the madness of pre-deployment staff preparations, inoculations, family pressures and sorting out a mound of green garb or naval combat gear, only served to sweeten the sense of escape as a half dozen of us from MARPAC, including Lieutenant-Commander Jamie MacKay (J4 Logistics (Log)), Lieutenant-Commander Chris Henderson (J5 Public Affairs (PA)) and Petty-Officer 1 Bruce Neuman (Yeoman), with a cold pack of Japanese encephalitis vaccine in hand, slipped away en route to Darwin the afternoon after HMCS *Protecteur's* emotional departure. That morning, 23 September, and three days after the INTERFET landing, the ship's role was still focused on humanitarian relief ashore, serving the JTF HQ and an element of refueling work, the potential conflicts of the latter two growing more apparent.

THE WAIT IN DARWIN

Australia is a continent far, far away. This fact would be brought home many times during our deployment, and for many reasons. Arriving there, we set to work to build our contacts and the network of offices that we knew would be vital in coordinating communications, port services, air transport, and the like. Darwin, it must be understood, is the warm, sandy, and often monsoonal equivalent of Churchill, Manitoba. It is still a frontier town, with limited jetty facilities, no rail head, roads subject to flooding, and limited infrastructure. In time, it would be paradise to Dili in East Timor, but I am getting ahead of myself.

Literally on arrival, our first meeting was in fact with our own Canadian Strategic Reconnaissance (Recce) Team, which had concluded its visits and even managed to parlay a short stop in the fledgling INTERFET HQ in Dili, a real coup resulting in some invaluable information and impressions. To my knowledge, no such Staff/Recce interface had occurred before, at least not at the outset of a mission such as this. We found the link invaluable and profited from it for the six months we were there. With INTERFET's lead elements now in place and the international force mix clarifying, the Australian Defence Force (ADF) requirements were starting to gel. Land forces for Dili were of less importance now that forces had been inserted. Talk of a "Commonwealth Brigade" was initiated. The need for an at-sea fueling capability was raised to some priority. The sense that INTERFET HQ was now the centre of information and decision-making was resoundingly clear.

From a purely military perspective, Darwin is relatively well endowed, the result of some politics to be sure, but mainly I think a purely natural consequence of having been bombarded several times by the Japanese during World War II. The ADF's Northern Command (NORCOM) has its headquarters there, and all three services are represented with facilities including airfields, barracks and support elements. NORCOM itself is a standing joint headquarters, headed at the time by Commodore Mark Bonser. Received with open arms, we were offered full access to the operational picture and the daily brief. Expecting something akin to the level of detail in the MARPAC or MARLANT morning briefs, what was available was at the strategic level only, and though Darwin's transport aircraft and troop count was listed, very little sense of events in East Timor, what we needed most, was presented, the occasional highlight aside. In time it became apparent that this more detailed picture was painted for the senior leadership of the ADF largely by phone calls and networking, with Major-General Peter Cosgrove himself often serving to pass the word direct from Dili.

What was immediately evident was that the true picture and the decision-makers, at least in terms of the likely operational employment of Canadian assets, were in Dili. The need to get to the coal face drove us to arrange a recce for 28 September. A team of us went over, including Captain Bruce Sand, my J3 Engineer, who would look at bed-down issues and Major Andre Corbould, my interim Chief of Staff (COS). Our aim was to tie in to the INTERFET staff and glean as much information as we could about the ADF concept of operations and where we might fit in. Understand that apart from the Airlift Task Force (ALTF) the Canadian commitment was still developmental, particularly the land element.

CHAPTER 2

As we flew in, the effects of the dry season were striking, with red ochre dust and dry vegetation prevalent everywhere except on the coast. The mountainous terrain and dried river beds were impressive, but it was the nooks and crags and canyons that struck home the potential dangers of a protracted guerilla war in this unfortunate country. Flying into Dili, a coastal town ringed by substantial hills, fires were still evident and a smoke haze was everywhere. Despite the madhouse of activity at the largely demolished airport, Major-General Cosgrove met us on arrival and confirmed a session for discussions later in the day. Troops and trucks and firepower were everywhere, but given that the heights were not yet secured, the vulnerability of INTERFET's toehold in East Timor was also evident. Militia were known to be present, although the town remained largely deserted. The militia's bounty on the INTERFET leadership demanded robust personal security. Even the TNI remained in large numbers, notionally restricted to their barracks, but they and the gunboats and submarines in the area made it clear that Indonesia was still interested in what was going on and that a turn for the worse could get very ugly very fast.

We moved by convoy to INTERFET HQ, a comparatively lightly damaged three story building that had merely been torched and had the widows and stair railings smashed. Commodore Jim Stapleton, the Naval Component Commander (NCC) was my first stop, and we discussed HMCS *Protecteur's* employment. He explained that HMAS *Success* had just spent six months assigned to their mission in Papua New Guinea, had been running very hard and was in need of a maintenance period. The likelihood was that HMCS *Protecteur* would be the only AOR available to the force for some time, and that maritime replenishment would rate high on her dance card from his perspective. We spoke of the games the two Indonesian submarines were playing and the tensions of having TNI landing ships and gunboats staring down the necks of INTERFET assets. The issue of shore support from a logistic and a combat perspective was a new twist, but one which he had not given much thought to yet. It was noted that at least one INTERFET Special Operations patrol had gotten in over its head and required extraction by a frigate. Of interest, the NCC was clearly an add-on element of the headquarters, as apparently was the Air Component Commander, with no air or naval staff included in the Combined (C)5 planning cell. This apparent oversight was not merely a function of the forming of an ad hoc HQ, but seemingly one of national doctrine and training. Clearly, ADF jointness was not jointness as we knew it in Canada.

The changing scenario of potential employment was compounded in discussions with the C5 Plans staff, who had assessed Dili as being made secure in the coming days and saw the focus of robust forces being required along the border areas to the west. A number of issues, including language, doctrine, interoperability, training, experience and ROE among them, were driving the Australians to consider placing two brigade groups made up of Commonwealth troops in this domain. A New Zealand/Canadian battalion was considered at this point, and while location was still undetermined, the south coast was coming into focus because of the level of militia activity in the town of Suai.

So much for the Dili-centric Canadians, and the relative comforts of a floating JTF HQ were in doubt as well. It was my discussion with Major-General Cosgrove, Commander of INTERFET, that afternoon that convinced me that the place to be was in Dili. His vision of the potential challenges and pace of activity were quite clear. Indeed, of all the Australians I met in theatre, he struck me as unique in his profound understanding of the Asian issues and complexities at play on this particular stage, and in his sensitivity to the aspect of "face" for the nations involved. His major concerns were logistics and the coalition's ability to deliver both force and sustainment to a country whose infrastructure was essentially wiped out, and perhaps more importantly, the robustness of the forces at his disposal, including from a ROE perspective. This later pertained, despite the robustness of a Chapter Seven mandate, to some participating nations having quite restrictive national ROE, including one that had an absolute restriction on deadly force. This reality would of course come to play on the eventual disposition of contingents, including our own.

A closing session with the Force Provost Marshall to discuss security matters conveniently led into a guided tour of Dili on our way back to the airport. Only a few thousand civilians were in a city which had housed about a hundred thousand. Apart from the occasional press-hired motorcycle or bashed UN Land Rover, the only vehicles were military. Apart from the absolutely mind-boggling destruction of the town, the other great incongruity was the cheek to jowl presence of thousands of TNI infantry, marines and special forces, some of whom were reported to have participated in the rampage of the previous weeks. The scene at the port was particularly memorable. The place was partly still alight, half remained occupied by the TNI. The few buildings abandoned so far for the use of INTERFET reeked of smoke, garbage and excrement. At the west end, refugees were crowded to standing room only in an old warehouse, unsure of whether to fear a resurgence of TNI/militia outrage or believe the disinformation about the "murdering INTERFET killers and rapists." This was my first lesson of many about psychological operations and the East Timorese.

By the time we boarded our Royal Australian Air Force (RAAF) Kingair aircraft for transportation to Darwin, we had been filled with tasks, plans and ideas. A busy but fruitful day had seen HMCS *Protecteur's* role change, our headquarters potentially move ashore and our land element potentially adjust both role and location. Not bad for a day's work, and justification to get staff into Dili in a full time capacity as soon as possible, to keep track of the evolving scene from within rather than from afar. Both the daily call to J Staff in National Defence Headquarters (NDHQ) and the situation reports (SITREP) were long ones. The changing picture of employment was affecting the plan and would in fact force adjustments to TO&E. Agreements for support with the ADF, and primarily the New Zealand Army, were negotiated and drawn up. The agreement with the New Zealand government evolved as a result of confirmation that the Canadian company group would form a significant portion of the 1st Battalion, Royal New Zealand Infantry Regiment (1 RNZIR) contingent, along with an Irish platoon and eventually a Fijian Company to round out the force. The logistics agreement was hoped to limit the material and personnel requirements for the land element. Time would tell.

Back in Darwin, we continued our efforts at preparing for the forces to come, particularly in sorting out the support side of things in Darwin. Of course, the ALTF was already on scene, having arrived on 28 September, and making a solid contribution. The bad press these folks received as a result of the glitches on their transit to Australia belies their true value to the mission. I was amazed at how they were up and running and on their first sortie within 24 hours of their arrival in Darwin. During their tenure with OP TOUCAN, they would carry 40 percent of the cargo with only 10 percent of the aircraft assets, an amazing feat. In time, the Royal 22nd Regiment (R22eR or "Van Doos") arrived in Townsville for acclimatization and training and HMCS *Protecteur* arrived in Darwin after her cross-Pacific trek and mission Work-Ups. With ship's preparations, sea lift for the troops, construction engineering plans, "national" options for airlift and many other issues on the go, things were clearly coming to a head. With inoculations complete, field kit and weapons in hand, at least figuratively, it was also time for the headquarters to make the shift from our ad hoc set-up in Darwin and install ourselves into the fray in Dili.

FINALLY IN DILI

The JTF Timor HQ was set up in the shell of what had been the East Timor Provincial Forestry and Agriculture Administration, a positive turn of events for us that the Yeoman managed to arrange through the deft use of his naval procurement skills. Neither Canadian nor Australian army ever understood just what a Yeoman was, but after seeing ours they all wanted one. Only slightly burnt out and with a comparatively sound roof, we started populating and cleaning up this two story building with the beginnings of a staff and two signalers.

I am sure there were a number of occasions on this land-focused operation where a naval officer's bliss was in fact ignorance, and setting up an HQ ashore was one of them. Having spent a dog watch or two in an operations room, I was more than comfortable with running a busy operation with a dozen people or so. The concept of being borne in the AOR where we could utilize the talents of Commanding Officer (CO) and crew fell away, but for a number of reasons the size of the staff element did not grow appreciably, and certainly not to the proportions that previous missions had come to enjoy. With about ten core staffers, another half-dozen in INTERFET HQ, a half dozen support personnel, a ten-person Military Police Detachment from MARLANT and a Kingston-based Signals MM Satellite Communications Detachment, we eventually maxed out at about forty-five personnel.

It will be a long time before images of sandbags, barbed wire, field showers, jerry cans, cots and "mozzie domes" or the sounds of barking dogs and roosters that light off at the rising of a bright star fade from memory. By the end of October, some fifty thousand people had returned to Dili, eventually to grow to one hundred thousand. Cars, mopeds and bicycles were everywhere and traffic signs or pedestrian control was not. Cooking fires, road dust, exhaust and the "mosquito fog wagon" assaulted the lungs. Temperatures of 45 degrees Celsius (C), 56°C on the tarmac at the airfield, and a mean humidity of 92 percent made for a demanding acclimatization, even after weeks in Darwin. After being lulled into a sense of security by a few light showers, our first true monsoon rain literally set us back on our heels, as we applied damage control training to plug holes and fend off the absolute torrents pouring into our office/dwelling. I have never in my life seen it rain so hard. An Ottawa Valley thunderstorm is a mere spring shower by comparison. So hard did it rain, and so well did the militias sabotage the sewer system, that within hours the city of Dili became a shallow bowl knee-deep in biology mercifully masked by sunset and the darkness of failed generators.

The crux of the matter in driving the decision on Dili was simply that INTERFET HQ was the source of information and the battle rhythm demanded senior participation a number of times during the day. Even if at anchor in the approaches, maintaining connectivity from HMCS *Protecteur* would have been difficult. With her away for days at a time refueling or supporting forces ashore, it would have been untenable. It is important to note that my own role as Contingent Commander entailed operational command (OPCOM) of Canadian forces, but operational control (OPCON) was delegated for each element to a foreign HQ. For example: HMCS *Protecteur* to the NCC, the ALTF to the Coalition Wing Commander, and the R22eR Company Group to the land force commander through the New Zealand battalion commander. This was very clear to all of us, Canadian, Australian and New Zealander, and was perhaps the key point made by both the Chief of the Defence Staff (CDS) and DCDS during my calls before departing Ottawa. While this meant that we were not responsible for day to day operations, the oversight role in ensuring that the assigned Canadians were employed appropriately, and the advocate or "godfather" role, each entailed volumes of work and constant communications.

Section base teams aside, it became clear how thin we were on the ground in Dili. Our designation as an "austere" mission had a number of implications, and control on establishment was one. With the many evolving steps in defining the land element and an eventual confirmation of the R22eR company group's relatively isolated position in Zumalai, priority was given to rounding out the land assets and their support, as opposed to growing the headquarters. While I would never regret that decision, the pace of activity for the staff was indeed extreme and by December even the J-Staff acknowledged that we had gone in particularly light. Managing the requirements of five units, including the HQ itself and the two cells of the National Support Element (NSE), located in six disparate locations proved to be a challenge of orchestral proportions, but fun nonetheless. The saving grace of course was the talented leaders which I had been provided to help keep watch over the fold and the trustful relationships which we quickly built.

The key was of course maintaining a network of connectivity. This entailed morning and evening sessions in INTERFET HQ, ongoing feedback from our INTERFET staffers, particularly Lieutenant-Colonel Rocky Lacroix in the C5 Plans cell, liaison with the units to sort out laments and recommendations, and a daily SITREP amplified by a daily call to Colonel Walt Natynczyk in the COS J3 shop. Communications-wise, we were extremely well supported by my J6, Captain Ken Mahon from Kingston, and his signals element, who provided national satellite connectivity for message, phones, satellite communications (SATCOM), Defence information network (DIN) access and internet, and liaised with the ADF and American communications support teams to put the coalition "Parakeet" field/satellite phone on a number of desks. In short, communications were not generally a concern, having the staff to sort the wheat from the chaff and effect a fix sometimes was. This latter was endemic throughout the coalition, and it was amusing at times to see a three day old report brought to a brief as new information by a section that had simply been cut out of the loop or which was too overworked to get to their traffic.

Throughout this early period in Dili, thousands of troops continued to pour into East Timor by sea and air, as did hardware, fuel and supplies. The lack of infrastructure, inadequacies of the port facilities, lack of lay-down space and movements equipment at both port and airfield all conspired to slow down and sometimes halt the flow. The lack of "movers" was such that these folks worked consistently 18 to 20-hour days and safety issues eventually overrode operational demands when injuries mounted. Military dependence on sea containers was irrelevant in a third-world country with no handling capacity, two small piers and a ten meter harbour depth. At one point, the bottleneck for materiel transport was an inability to build enough pallets to load the transport that was available. Fortune did smile in the form of the leased "fast-cat," HMAS *Jervis Bay*, which had just enough "legs" to make Darwin to Dili and return, carrying troops, kit and light vehicles.

Another impediment to the support network that became apparent was the way the support staffers were left out of the planning process.

If INTERFET HQ was army oriented, the C5 plans cell was small and purely land-op focused, sometimes surprising C4 (Log) or the NCC with sudden demands or changes in priority that were at least painful, and at times almost impossible. The changing nature of the situation aside, the exclusion of support staff from the planning process made resource planning and management a nightmare. As I have already said, the logistics war was a close run thing for the coalition. We must always remind ourselves that we do not do operations without support, and it is a tribute to the many staff and support personnel involved in both Darwin and East Timor that the many challenges were overcome.

From the Canadian perspective, our focus of activity was the imperative of establishing the R22eR company group ashore, first in Suai and eventually in Zumalai. Both HMCS *Protecteur* and the ALTF were well ensconced in their routines with NCC and Wing in short order. Having overcome the hurdle of simply getting the troops from Darwin and across the beach, their next challenges, the issues of support, transport and living conditions quickly arose. Staffing of support elements and equipment, initially lean because of the ADF and New Zealand support agreements, became an evolutionary growth that would last through January. Transportation issues had to meet the challenges of a 10,000 square kilometres (km) area of operations, poor roads and cattle tracks, as well as a climate which meant a soldier could only carry enough water to go about 5 km in a day. Here the solution was all terrain vehicles in the form of a small off-road 4x4, and a fun ride they were. Moreover they were effective and safe and a delightful curiosity when trying to communicate in the small isolated villages.

Living conditions and accommodations became a major concern for us. The realities of the tropics, and of the monsoon in particular, made it clear that austere or not, hoochies were going to be inadequate. Major Pat Heffernan, from 1 Construction Engineering Unit (CEU), surveyed the area and assessed the climate, delivering a business case that eventually translated into some 29 steel "meccano set" buildings – the Canadian jungle hut, at a cost of some $700K. We earned the envy of our allies and the wonder of the people of Zumalai as 1CEU, the company group field engineers and MARPAC's Naval Constructor Troop transformed a soccer pitch into a miniature town using field equipment, bobcats and what had to be a "BINFORD" toolbox.

To me, the result was a force multiplier. Instead of a daily sick parade of 25-30 soldiers, we dropped down to 5-8. Meanwhile, months later, our New Zealand and Australian counterparts were still living in hoochies and sea containers and using "outdoor" ablution facilities. Given the value of the training and expertise of our personnel, and if we stopped just ten soldiers from putting in their releases as a result of not having to face what would have been hellish living conditions, then we paid for the camp. We made the right call and the Canadian Forces (CF) Engineering world have much to be proud of in what will be our legacy in Zumalai.

FLEETING STEADY-STATE

Steady state is an elusive beast with a one-off mission such as OP TOUCAN. Though we looked at a possible rotation scenario early on, and while no firm decision was delivered by Government till early February, it was reasonably clear even in December that we were it. Discussions of disengagement and recommendations for follow-on contributions had been on the go for weeks at that point, and many contingencies were put together. All we required was some clarity as to national policy and a confirmation of direction.

In the meantime, the CF contribution was in place and fully engaged in various operations, Camp Maple Leaf was commissioned and the network of daily communications well established. HMCS *Protecteur* was serving essentially as the sole source of petroleum, oil and lubricants (POL) products at sea, as well as for forces ashore. This was a potential single point of failure that could have brought the entire coalition to a grinding halt, air-delivered fuel notwithstanding. The R22eR company group, with veterans of Bosnia, Kosovo and Haiti, set an example for thoroughness and fairness and delivered a number of thought provoking recommendations and observations to their New Zealand battalion HQ through coalition headquarters, including somewhat ironically, work on the issue of dealing with detainees.

Living conditions in Zumalai, though never comfortable in the muck of the monsoon, were comparatively enjoyable and a distinct source of pride. With an element of routine setting in, at least for the field units and for HMCS *Protecteur*, humanitarian projects took on a level of priority and, with cease operations on the horizon, some sense of urgency as well. This effort in particular was magnificent to see, as churches, schools and hospitals were re-roofed, refurbished or re-built. "Christmas KISS" and other projects delivered teddy-bears, parcels and school supplies. An ancient tractor was re-built essentially from scratch. Water works, community power and pumping stations were improved. Language tutors, some utilizing Valcartier English, sprung up. Despite the backdrop of destruction and death, the sense of a hopeful future, and the inspiration of a warm and caring people, helped carry our troops through a tough deployment.

POSSIBLE SCENARIOS

I will always say that the tools that Canada brought to the workshop that was East Timor were appropriate and effective, because they truly were. That said, it is impossible to avoid reflecting what might have happened in a perfect world, because missions such as these have and will come our way with some frequency. Operational logistics support, heavy lift helicopters and littoral sea-lift sum up the three areas where I believe we were most vulnerable as an international force. Yes, combat capable forces in sufficient numbers were a concern from time to time and in certain locales, but these three were consistent and abiding concerns. I termed the latter carefully because "amphibious" conjures up images of D-Day (i.e. the Normandy invasion in World War II) and such offensive activity or the casualties it represents mires the term in automatic resistance. What I am referring to is a troop and equipment-carrying sea-going platform that can get personnel and equipment into a port or across a beach. The beach aspect is vital if we expect disaster ridden nations, natural or man-made, to simply cater to RO-RO (roll on – roll off) capabilities, which in this case HMAS *Tobruk* and the Fijian Navy Ship (FNS) *Siroco* catered to with a great degree of stress-relieving success.

While HMCS *Protecteur* did a marvelous job, I wonder how we might have had more control over our own national movements and sustainment, as well as what we may have been able to achieve as a force from a coalition perspective, and the kind of impression Canada might have been able to make as a nation, had ALSC been made available, fueling capability and all, to me or to the NCC. Hopefully in time we will know.

The helicopter issue is interesting, especially given that the solution came in the form of American Seaknight helicopters, which look amazingly similar to the venerable Chinook helicopter. Whether Canada will play in this game ever again is debatable, but the rarity of this important platform in the international air community is significant. A niche perhaps? A tie-in to offshore search and rescue (SAR) work or transport in the Canadian Arctic?

And last the logistics issue, which in my opinion was the force's most vulnerable aspect for so many reasons. These include the limitations of the airhead and port facilities in both Darwin and Dili, and the limitations of transportation capacity and movements expertise in general for several months. The fact that the Australian Army is by doctrine a continental army not accustomed to deploying afar is a fact that led to reorganizations and restructuring of the force, which was stretched to the breaking point in providing for a division-sized expeditionary force some 200 miles from Australia, 24 hours by sea and 90 minutes by air. That some recent experience had been gained through their recent operations is more fate than conscious planning, but fortunate nonetheless. What I still find amazing is that neither the TNI nor the militias seemed to sense or exploit a vulnerability that had the potential to bring the entire operation to a screeching halt. Our own expertise in this field, with both NATO and UN roots, is renowned and justifiably so, as I witnessed with the teams we brought into theatre.

WE LEARNED A LESSON OR TWO

A number of lessons learned arise from any deployment, and OP TOUCAN is no different. I will not detail the minutiae of the post deployment report, but two points come to mind in particular. First, that logistics support has critical mass and that the expectations of an ally's efforts may be optimistic, particularly if allies are themselves stretched thin. From what I have observed in the context of OP TOUCAN, we do logistics exceedingly well in Canada, a capability worth holding on to if we are to continue to traipse all over the world as we have done now for decades.

Second, an armada is still an impressive and intimidating sight. According to Major-General Cosgrove himself, the coalition's amassed tonnage and naval might in the approaches to Dili helped convince the TNI and the Government of Indonesia that the international community had in fact ranged itself in full support of an independent East Timor, in a way that even his forces ashore could not. Sea power as a diplomatic force is alive and well.

WHAT THE NAVAL OFFICER BRINGS

Before wrapping up, I want to address the issue of what a naval officer brings to the playing field during deployed operations such as OP TOUCAN. The issue is not how we apply our specialty at sea or what we add to our portfolio by doing something unique, but what it is in the naval officer's inventory in general that can enhance the mission. While acknowledging some of the unique attributes of OP TOUCAN, and noting that ours was no more perfect a mission than any that had gone before, I found time and again that the training, experience and background that I had been fortunate enough to amass over the years served me exceptionally well, not necessarily to predict challenges or foresee problems in this new environment, but certainly to overcome them. With more than my share of time to mull over the issue and to observe various players of all ranks, backgrounds and nationalities at hand in INTERFET, what became clear, not surprisingly, is that we are indeed products of our environment. The revelation to me was how well the naval service did in Timor, ashore and afloat, and particularly in a joint and combined staff environment. Noting that the Air Force thought load and sortie, while the Army seemed to focus on task or objective, I considered, I think for the first time, how it is a maritime surface (MARS) officer thinks of things. I concluded that we tend not to think of one place or task or serial or even mission, that after our CCO or ORO course we have our circuitry adjusted to think "voyage". This means point A to point B over time and in three dimensions, with the dynamic inputs that weather, consorts or the enemy can bring, with a mindset of the importance of planning, flexible responses and the ability to make it up on the fly. Not a bad skill set to have, and one that I came to find is in high demand. If the more junior staff officers which we had ashore are any measure, we continue to grow these skills and thought processes reasonably well today.

CLOSING OUT

The remaining elements of OP TOUCAN, that being the JTF HQ, NSE and R22eR company group, transitioned to the United Nations Transitional Administration East Timor (UNTAET) and the UN peacekeeping mission on 23 February, as marked by an international ceremony in Suai. OP TOUCAN's cease operations was represented by a small Canadian ceremony in Zumalai on 12 March, as Lieutenant-General De Los Santos, the UN peacekeeping force commander, received the

UN flag from Major Alain Gautier, the R22eR company commander. The short burst of energy that was OP TOUCAN was, I feel, an appropriate, flexible and effective force during a period of intense turmoil and potential risk in a classic "peacemaking" role. It was too much, too "pointy-end" and robust I think, sustainability issues aside, for the long period of peacekeeping that yet lies ahead. We have left behind three staff officers in UNTAET as Canada's current military contribution to efforts in East Timor, who along with a diplomatic mission, the Royal Canadian Mounted Police (RCMP), UN staffers and humanitarian aid agencies provide Canada's ongoing commitment to the region.

Our contribution was evident in the quantum leap with which living conditions improved in East Timor, and in the eyes of our gentle hosts when we said good-bye. By chance, during my last walk around Zumalai, I met a gentleman whom I had met months before when we first flew in by Sea King helicopter, and whom I had seen at the refugee drop-off in town on several occasions desperately seeking his family, which had been forcibly trucked off to West Timor, without success. This time he introduced me to his wife and pointed out his very shy children in front of the house he had rebuilt with contributions from our troops and sailors. We still had trouble communicating, but his eyes said volumes. When Canadians lament that we cannot change the world, I will always remember that we can, one step at a time. This level of satisfaction is only increased by the privilege of working with the absolutely superb professionals that I had assigned to me in OP TOUCAN – sailor, soldier, aircrew, maintainer, support trade, sapper, staff or whatever. In the end, it's hard to believe how lucky we are to have such challenges foist upon us, and be in turn blessed with such fine people to make it all work. Lucky indeed.

CHAPTER 3

IN HARM'S WAY: THE 3rd BATTALION, THE ROYAL CANADIAN REGIMENT ON OPERATION PALLADIUM, ROTO 3 – JULY 1998 TO JANUARY 1999

Colonel Mike Jorgensen

This account of the experiences of the 3rd Battalion, the Royal Canadian Regiment (3 RCR) Battle Group (BG), on Operation (OP) PALLADIUM, Rotation (ROTO) 3, in support of the General Framework Agreement for Peace (GFAP) aspects of the "Dayton Agreement" has been prepared in support of the *In Harm's Way* series. In preparing the account, I have generally followed the five doctrinal phases of peace support operations to describe the experiences of the Battle Group – warning, mounting (preparation), deployment, employment, and re-deployment. However, the major portion of this chapter is dedicated to the "employment" phase, where we had a great opportunity to put our professional abilities to the test.

I have chosen to focus on a number of very specific events that occurred in our tour – key events that provided the members of the Battle Group an opportunity to fully employ their professional skills. These key events include support to the elections, OP SHANNON, and our support to rebuilding efforts, all of which involved all members of the Battle Group and most of the Canadian Contingent.

WARNING

The leadership of the 2nd Canadian Mechanized Brigade Group (2 CMBG) had gathered late in the fall at the home of the Brigade G4 (logistics), Major Ryan Jestin, as directed by the Brigade Commander, Brigadier-General R.J. Hillier. Hillier was returning from National Defence Headquarters (NDHQ) after having received direction with respect to the Army's choice of lead unit, manning limits, and so forth, in support of OP PALLADIUM ROTO 3. Upon arrival at the Jestins' married quarter, Brigadier-General Hillier briefed the unit commanding officers on his concept of operations for the mounting and deployment of the 3 RCR BG. We received our formal warning order on 7 November 1997.

The BG leadership paid a visit to the Deputy Chief of the Defence Staff (DCDS) Joint (J)-staff in Ottawa early in our planning phase, and later hosted their team in Petawawa. We mostly received our written, formal direction from the DCDS J3 International Staff, through the chain of command. Most of the directives arrived late, or had been overtaken by events by the time they arrived. In our case, Land Forces Central Area (LFCA) HQ had the lead for force generation and mounting, while 2 CMBG was tasked to train and declare the unit operationally ready. The commitment from 2 CMBG – making our preparatory training a Brigade main effort – was a key contributor to mission success.

As a unit, we immediately developed an overarching and comprehensive plan, and established a formal "writing plan" to capture and communicate all aspects thereof in a centralized manner. We planned in close consultation with the Brigade HQ, and other units within the Brigade who were impacted by the tasking. Later in the planning process, we began to communicate more frequently with the 1 RCR BG, which was in theatre. The "writing plan" proved to be a very effective mechanism for capturing all facets of the operation. The Deputy Commanding Officer, Major Buster Bowes, shepherded this project for us. The plan served to initiate the detailed planning for the operation – including every necessary piece of documentation we would ultimately need to produce in order to effect all phases of the operation. Ultimately, we filled three large binders prior to deployment, and a further two binders in theatre.

Upon receipt of warning orders and operational and administrative guidance, we embarked on personnel selection, screening, and preparation. This is something commanders at all levels must become personally engaged in from the start, with a view to ensuring, to the degree possible, that personnel with administrative, personal, or medical problems, are not included in the deployed component of the force. In my view, a poor screening effort will ensure that these individuals eventually create significant administrative or disciplinary problems for you and your subordinate commanders. Our Regimental Sergeant-Major (RSM), Chief Warrant Officer Frank Grattan, played a huge role in this process, and he and I personally interviewed Reservist Officers and senior non-commissioned officers (NCOs) who had applied to join the Battle Group.

Another preparatory activity that paid off nicely was the conduct of an effective in-theatre reconnaissance. In fact, we conducted two of these – to great effect. The first of these was the "Commanders' Reconnaissance," which included our brigade commander (Brigadier-General Hillier), the Commanding Officer (CO) of the "Training Assistance Unit" (i.e. the CO, the Royal Canadian Dragoons (RCD), Lieutenant-Colonel W.J. Natynczyk), the 2 CMBG HQ G3 (Major Bruce Pennington), and the 2 CMBG HQ G4 (Major Ryan Jestin). The Commander's Reconnaissance allowed us to collectively plan our pre-deployment training concept. The other, more traditional reconnaissance, was conducted with sub-unit commanders and key unit staff officers, which allowed for the development of detailed technical plans in anticipation of our handover.

Instituting a dynamic and effective rear party organization played a major role in creating calm and confidence in our pre-deployment activities, and served us superbly while deployed. Major Kevin Robinson, ably assisted by Master-Warrant Officer Don Campbell, commanded the Rear Party. It included an energetic spouses committee, a support network, and a comprehensive fan-out system. The Military Family Resources Centre supported our efforts, and provided additional useful information and contacts for this critical component of our team.

We began planning for leave travel assistance trips early, under the Welfare Officer – in this case, Captain Will Graydon. Early planning and decision-making is important in this quality of life aspect of the plan. And, equally important, is the requirement to ensure the plan is characterized by a degree of "consensus", transparency, and "buy-in". Communicating the plan early, during "family briefing" nights, was also a very helpful technique, serving to promote high morale and family confidence.

MOUNTING (PREPARATION)

We conducted our pre-deployment preparations and training during the period 5 December 1997 to 4 June 1998. We had an official "stand-up" parade in the new 3 RCR compound on 5 December 1997, presided over by the 2 CMBG brigade commander. Shortly afterwards, the newly formed Battle Group departed on Christmas Leave. However, the Battle Group was recalled from leave on 8 January 1998, in order to participate in OP RECUPERATION – the "ice storm of the century"! The brigade commander, key Brigade staff, and the unit COs departed for the National

Capital Region (NCR) at 0100 hours, on 9 January. In fact, we spent almost three weeks working in support of OP RECUPERATION, leading me to approach the brigade commander with a view to having us return to Canadian Forces Base (CFB) Petawawa first among the Brigade units, given our urgent requirement to undertake a full slate of Primary Combat Function courses as a first step in our pre-deployment training package. In fact, the "ice storm" deployment allowed the newly formed Battle Group to begin the "bonding process" – less the armour squadron, which reverted to "under command" of the CO of the RCD for the Operation. Our BG soccer team later returned to play against a local team in the Alexandria area, returning from there with the "Ice Storm Cup"!

In the end, our pre-deployment training program began approximately three weeks after the intended start date as a result of our deployment on "Operation Ice Storm". The program began with a series of individual training courses, beginning at the end of January 1998, and continued until the unit went on leave after the Brigade Commander declared us operationally ready at a Battle Group parade conducted on 4 June 1998. The role of the higher HQ – in our case 2 CMBG HQ, was defined by the brigade commander, Brigadier-General Hillier as "setting the conditions for the success of the 3 RCR BG". In turn, our pre-deployment efforts became the main effort of the entire formation, which turned out to be exactly what was needed.

Our training program proceeded very logically from general-purpose combat training to mission-specific training, working from individual level to collective level throughout. I was struck by the requirement to institute a period of transition from one to the other, as the shift in mind-set is important for mission success – although losing the combat focus entirely is equally dangerous. We also sought to ensure that we would have sufficient depth in terms of individual qualifications (i.e. primary combat functions) to be able to react to attrition, promotions, medical problems, leave, and so forth. Accordingly, we conducted individual training during the period 12 January to 19 February 1998, including driver (wheeled, tracked, and Armoured Vehicle General Purpose (AVGP)), machine gunner, and TOW (Tube launched, Optically tracked, Wire guided missile) gunner.

The Brigade Commander designated the RCD as our "training assistance unit". We had previously assisted 1 RCR in its preparations for a deployment to Kosovo and Bosnia, so we were very familiar with the concept of a "training assistance unit". The designation of a "training assistance unit" was very useful for our purposes. It allowed our "teams" to focus on training as cohesive elements, uniformly and conveniently. In our case, given that we had an excellent rapport with our "training assistance unit", and we enjoyed top-notch support from day one, we were able to focus on developing the required skills rather than spending our time analyzing the training objectives, coordinating training support, or providing instructors or role-players.

Our pre-deployment collective training phase was planned for the late-winter period. In fact, had we stayed in Petawawa, we would not have been able to complete any worthwhile manoeuvre training as a result of the weather (snow and ice). In our case, one of the more inspired decisions was to allow us to travel to Fort Chaffee, Arkansas, to complete a large portion of our collective training there during the period 23 February to 10 March 1998. The entire process of planning and completing the 1,700 kilometre road move, and the associated air lift, provided enormous experience to our administrative component – while injecting an air of excitement and novelty into the training schedule. Most importantly, it allowed us to practice manoeuvring our combat teams in appropriate (but very challenging) terrain while snow covered our home base in Petawawa. The entire Canadian contingent (with a few exceptions, including the National Command Element (NCE)) deployed to Fort Chaffee, which allowed us to "bond" as a contingent and reinforce lasting relationships between the various commanders.

The deployment to Fort Chaffee included a Coyote-equipped opposition force (OPFOR), led by Major Jeff Barr, who provided outstanding training support to the Battle Group. The confirmation part of the plan suffered somewhat at the absence of Brigadier-General Hillier who was prevented from joining us because of the unfortunate sudden passing away of his father. The National Support Element (NSE) completed a series of individual training requirements while in Fort Chaffee, and then supported our training program administratively.

Our road move became legend immediately! Our rail-move skills were markedly improved at the end of the move, and our vehicles received an excellent road test. Our Advance Party – which included Captain Steve Nash, had established an excellent rapport with the Fort's facilities personnel, and we received several distinguished visitors, including the Adjutant-General of the Arkansas National Guard. The final ("confirmatory") exercise, coordinated by the G3 and G4 of 2 CMBG, provided several exciting moments, including a final attack at Battle Group level.

Aside from the general-purpose combat training we completed up front, we followed a fairly standard mission-oriented training package – organized for us by our training assistance unit in consultation with the Battle Group leadership, and designed to provide us with a rudimentary background to the environmental, cultural, historical, military, and political situation in theatre. We spent considerable time training on the rules of engagement (ROE) – in fact, they were part of every single training activity we undertook, starting with a highly interactive "War Cabinet" training session. Training on the ROE was likely the most sensitive aspect of our training program, and we made sure we benefited from a wide variety of techniques, including lectures from the military lawyers serving on the J-Staff and round-robin "stand training." At one point, I personally summarized the ROE before the members of the BG, to lend a sense of confidence in the process – both in terms of confidence in the ROE and in terms of my own sense of comfort with them. We worked up our ability to "manage operations" through various means, including a large-scale Command Post Exercise (CPX), conducted in the 1 RCR unit lines.

Importantly, a large part of our final training phase – the mission-specific training phase, was completed on private land "cleared" for such purposes. We enjoyed tremendous support from the local community in this respect, which allowed us to conduct our training in urban areas to enhance training realism and relevance. The confirmation exercise, Exercise (Ex) Storming Bear II, took place in the area surrounding the NCR, partly to exploit the goodwill established in the Region by 2 CMBG during the "ice storm," during the period 25-29 May 1998. At the outset, our Signals Platoon experienced some real challenges in providing area-wide coverage, given the atmospheric clutter provided by the incredible concentration of signal-emitting apparatus! They had a mini-retreat on the spot, under the guidance of their "Atomic Engineering" qualified platoon commander, and shortly thereafter their brainstorming session allowed them to provide the necessary communications support to the exercise.

Of particular note was the close supervision exercised by the chain of command on the final day of the "confirmation exercise" – I recall it vividly, even today. The final "problem" involved a "holed-up" force of well-armed and desperate rogue Entity Armed Forces (EAF). Major Mike Evans was the company commander on the spot, and he had his platoons formed up around the objective, while he was attempting to "negotiate" with the folks in the buildings. As he was supervising his platoons, I was providing direction to him, and coordinating the supporting assets that were just arriving – including helicopters (portraying helicopter gunships), Military Police (setting up a Prisoner of War cage), and so forth. However, supervising me – from a distance of less than 20 metres, was (all within 10 metres of each other) the Brigade Commander, the Commander LFCA (overall responsible for a declaration of readiness) – Brigadier-General Walt Holmes, and the Commander of the Army – Lieutenant-General Bill Leach. All the Formation and Command chief warrant officers were there as well. Fortunately, all went well, as the rogue Entity Armed Forces were quickly persuaded to surrender in the face of overmatching combat power.

Throughout the process, it became very clear to me that a sound communications plan for internal and external purposes, is critical. It permits wide anticipation, reduces the potential for misunderstandings, and keeps morale high. Regular briefings – to families, to the chain of command, and the community, are a key component of this process.

An initial visit and information session with the Military Family Resource Centre was highly productive. I found that our early visit to the National Defence Coordination Centre (NDCC), the National Defence Intelligence Centre (NDIC), and the DCDS/J-Staff, was invaluable. Similarly, we benefited from a visit from the Army Lessons Learned Centre. The issuance of clothing and special equipment provided our sole source of frustration – the process was incredibly cumbersome and consumed considerable time and resources.

DEPLOYMENT

Our deployment unfolded smoothly under control of the National Defence Movement Control Centre, on behalf of the DCDS. It saw us moving through the designated Airfield Port Of Embarkation at CFB

Trenton, and then flying directly to Zagreb, Croatia, the Designated Airfield Port of Disembarkation.

Upon arrival in theatre, we became a DCDS unit, and reported on a daily basis to two masters – the National Contingent Commander (NCC), Colonel W.J. Natynczyk, and the commander of Multi-National Division (South-West) (MND (SW)) – Major-General (UK) Cedric Delves, who exercised operational control over us. National operational command of the 3 RCR BG continued to be vested in the Canadian contingent commander, who reported daily to the DCDS, and focused on coordinating matters of a national nature, including the overall aspects of good order and discipline for all Canadians in theatre.

EMPLOYMENT

Arrival and handover

Upon arrival, we followed standard handover procedures. The Board of Inquiry and related briefings set the tone and provided the background for the handover from the 1 RCR BG to our Battle Group. Touring the area of responsibility (AOR) in a carefully coordinated manner, in the company of our predecessors, allowed for a comprehensive handover. I personally found it helpful to meet my international counter-parts, including the EAF commanders.

On a well-established mission, Contingent Standing Orders, Force SOPs, and similar critical direction are well-established. However, on our first trip to HQ MND (SW) we took the time to review several tactical procedures together, such as information collection and dissemination, security, incident handling, use of force (with a special eye to Canadian sensitivities related to dealing with civilian outbreaks of violence), and contingency planning related to heightened states of alert, mobile reserve employment outside the AOR.

The Campaign Plan and the Battle Group Mission

We conducted operations within our AOR in a manner consistent with the Stabilization Force (SFOR) Campaign Plan. The SFOR Campaign Plan was commonly expressed in terms of a series of supporting pillars, with designated lead agencies for each pillar of the plan as follows, from which our own operational framework flowed:

- Secure environment (SFOR)

- Economic recovery (Office of the High Representative (OHR))

- De-mining (UN Mine Action Committee)

- Police (International Police Task Force – IPTF, and the International Criminal Tribunal for the former republic of Yugoslavia – ICTY)

- Displaced Persons and Refugees (DPRE) (UN High Commission for Refugees (UNHCR))

- Elections (Organization for Security and Cooperation in Europe – OSCE)

- Arms Control (OSCE)

- Common Institutions (OHR)

We had used our in-theatre orders for the confirmatory exercise, as well as all preliminary exercises. This turned out to be highly effective and efficient, although operational security must always be the overriding consideration. In our case, using the "real" orders paid dividends because everyone had a thorough grasp of their mission, role, and tasks immediately upon arrival, and felt entirely comfortable in the AOR based on a high degree of situational awareness. As always, mission tasking breeds pride of ownership, initiative, and confidence; accordingly, I put maximum effort into casting my direction in terms of commander's intent, concept of operations, main effort, and end-state. Thus, these orders were well known to all members of the Battle Group:

> The 3 RCR Battle Group will enforce the military aspects of GFAP, while providing support to the civil aspects, within resources, in order to provide a secure environment for continued development of peace.

I shall achieve the intent of the Commander MND (SW) through an active programme of framework patrolling, cantonment and GOF inspections, and monitoring of EAF training and movement. To effect the required coordination with all agencies involved in nation-building we will employ a liaison cell dedicated to the full spectrum of civil and military liaison activities. We will take every opportunity to provide a clear and unequivocal demonstration of our combat power, and execute all operations in a resolute manner.

To provide a dominant presence to assist in the provision of a secure environment, I will deploy rifle companies to dispersed locations throughout the AOR, in the process assigning them responsibility for specific portions of our AOR. From dispersed bases of operation, the companies will accomplish BG-assigned tasks, including framework tasks and inspection and monitoring tasks. I will superimpose the reconnaissance squadron on the complete AOR to ensure total area coverage, with special emphasis on potential gaps amongst the rifle companies. I will employ the Anti-Armour Platoon to provide overwatch throughout the AOR, with attention to our flanks. Underpinning our efforts will be the ready provision of a wide range of firepower, including our integral Mortar Platoon and OAS coordinated by our centralized, highly mobile engineer assets, capable of reacting quickly to all incidents, including those involving UXO or mines. Our ability to respond will be constant with ready reserves at all levels, while contributing concurrently to the ability of MND (SW) to respond through our provision of a company-sized reserve. A sound and flexible system of first line CSS [combat service support] will allow us to effect sustained, decentralized operations to meet all exigencies.

Our support to civil aspects of the GFAP will be very much tied to the availability of resources and local circumstances. Our civil support efforts will be carefully scrutinized and closely linked to the conduct of general operations within the AOR.

Our operational focus will concentrate on: the return of DPREs [Displaced Persons, Refugees and Evacuees]; monitoring of the implementation of the election results of 1997; supporting elections scheduled for September 1998; promoting ongoing initiatives such as Op Harvest; and, conducting an aggressive in-theatre training programme to maintain our combat capability and provide a visible deterrence to a resumption of hostilities or new threats to peace. In general, I intend to consolidate the efforts of previous contingents and promote a climate in which the process can continue.

Our main effort will be directed at maintaining and verifying complete compliance of all EAF with the military aspects of the GFAP. I consider the EAF to be the centre of gravity in operations in our AOR. The full range of our capabilities will be devoted to our main effort, which we will discharge in a firm, fair, and professional manner as an all arms unit, proud to be members of a multi-national team.

My end-state foresees a secure environment within which agencies charged with implementing the civil aspects of the GFAP are affecting the desired progress with confidence. Our efforts will be seen as productive in establishing strong working relationships with all agencies in the AOR, and providing a critical source of coordination in support of these agencies. Within our AOR our legacy will be characterized by tangible examples of our successful mission, including both a growing confidence of local civilian institutions and highly visible CIMIC or civilian aid projects. Our re-deployment to Canada will see the safe return of all our soldiers, reflecting a successful force protection posture throughout the operation, following a perfect, seamless handover to the 2 RCR Battle Group.

Battle Group structure, tasks, and battle rhythm

The 3 RCR Battle Group numbered 883 all ranks, and we had approximately 300 vehicles of various types – including a large number of wheeled SMP vehicles. The Battle Group was composed of three mechanized/motorized rifle companies equipped with "Grizzly" AVGPs, an Armoured Reconnaissance Squadron of the RCD equipped with "Cougar" AVGPs, a Combat Support Company manned primarily by artillery personnel from the 2nd Regiment, Royal Canadian Horse Artillery

CHAPTER 3   IN HARM'S WAY

(2 RCHA), and an Engineer Squadron from the 2nd Regiment, the Canadian Engineer Regiment (CER) equipped with a large variety of engineer vehicles and equipment.

The Battle Group leadership was structured along traditional lines, with a Deputy Commanding Officer from 3 RCR (Major "Buster" Bowes), a RSM from 3 RCR (Frank Gratten), and the Operations Officer (Major Dave Berry) and the Adjutant (Captain Geoff Parker) both coming from 3 RCR. The Administration Company worked under Major Randy Kemp (3 RCR), while the three rifle companies were commanded by officers from 3 RCR – Major Dan Jakubiek (Parachute Company), Major Rob Walker (November Company), and Major Mike Evans (Oscar Company). Major Mike Nixon (RCD) commanded "A" (Reconnaissance) Squadron, while Major Nak Yun Paik (2 CER) commanded 24 (Engineer) Field Squadron, and Major Dave MacPherson (2 RCHA) commanded the Combat Support Company. Major Craig Dalton worked as the BG CIMIC Officer, leading a small, but key Battle Group team.

Our AOR was located in the North-West corner of Bosnia, spanning an area similar in size to Prince Edward Island. Our Battle Group Headquarters was located at Camp Holopino (Coralici), which also housed Oscar Company, most of Combat Support Company, and a large proportion of the Administration Company – including the Advanced Surgical Center of our Battle Group. Camp Maple Leaf (Zgon) housed Parachute Company, the Reconnaissance Squadron, and the Engineer Squadron. The Medical Platoon/Unit Medical Station was established at Zgon. The Camp at Drvar was occupied by November Company. We had established smaller outposts and patrol bases at various locations, including the town of Bihac, Bos Petrovac, and Bos Grahovo.

The very size of the AOR led to a requirement for two Radio Re-Broadcast (RRB) stations (Mount Gola and Mount Raz) to support communications across the AOR. Both RRBs were located on mountain-tops, and were extremely difficult to access. However, these two RRBs were critical to our communications ability, and we were lucky that both were commanded by extraordinary junior NCOs who ensured the RRBs were functioning fully and effectively throughout our tour.

The rifle companies focused on a broad range of tasks. These tasks included maintaining a visible presence through mounted and dismounted patrols (overt, but occasionally covert operations); conducting cantonment site inspections; monitoring training and movement of the EAF; establishing Observation Posts (OPs) as required; conducting liaison with local force commanders; maintaining camp security; supporting international organizations such as OSCE, IPTF, UNHCR and OHR; supporting the Public Information Operations campaign; and, completing one "visible community project". All companies maintained a Personnel Designated Special Status (PDSS) registry, and local Quick Reaction Forces (QRF), and all were prepared to execute portions of Battle Group Contingency Plans. The Reconnaissance Squadron focused on AOR-wide patrols and mobile observation posts, while the Engineers concentrated on supporting de-mining efforts, providing Emergency Response Teams (ERT), and maintaining the "cleared route trace". The Combat Support Company focused on coordinating Battle Group indirect fire and aviation support requirements, and coordinating combat power demonstrations as assigned by MND (SW). Senior Liaison Officers, Public Information Officers, Public Affairs Officers, and Visits Officers played key roles as well throughout the tour.

Our CIMIC efforts were especially noteworthy, with this aspect of our mission focused on promotion of the civil development agenda where possible. This entailed close cooperation with donors and donor agencies such as the Canadian International Development Agency (CIDA) and the British Department For International Development (DFID), a UK version of CIDA, as well as OXFAM, and many other agencies from Italy, Norway, Germany, and the United States (U.S.). The CIMIC section also identified DPRE success areas and supported the exploitation of these wherever possible.

Our weekly routine evolved quickly to having the command team and the broader Allied community gather on Saturday mornings in Coralici to review past and future operations of interest to all international parties operating within the AOR. The afternoon provided an opportunity for the "War Cabinet" to discuss internal issues and issues of long-term interest, and consider more fundamental issues "in camera". Sunday mornings became "Sunday routine", with some opportunities for individual soldiers to take a short break from the numbing routine.

A key part of our weekly cycle involved the "RISTA" (Reconnaissance Intelligence Surveillance and Target Acquisition) conferences conducted one evening early each week. Here we assembled key organic and Allied intelligence and information gatherers with key operations and plans staffs, with a view to collectively establishing Named Areas of Interest (NAI), and identifying information requirements (IRs) and Priority Information Requirements (PIRs). Then, armed with these NAIs, IRs, and PIRs, we allocated them in matrix format to the folks in attendance. In addition, in put them on the patrol tables for our AOR. The following week we would review our progress, and refine the process to suit our requirements. In this manner, we were able to exploit the work of the Joint Command Observer (JCO) teams, the Allied Military Intelligence Battalion (AMIB), the Contingent Commander's Advisory Team (CCAT), the "Communications Support" (i.e. electronic warfare) section, and a host of other disparate organizations.

Sometimes, in the conduct of routine operations, unusual situations would arise. Such was the case of the explosion at the Bosnian Ammunition Compound mid-way through the tour. Major Mike Evans had built up a solid relationship with the local EAF commanders in his area. This relationship led them to contact him immediately following the explosion of a significantly large ammunition storage compound, with a view to having him deploy some of his armoured personnel carriers (APCs) into the compound in order to rescue wounded Bosniac soldiers, or recover dead bodies. With my support, Major Evans deployed with several vehicles into the still-smoldering compound to support the local EAF commanders in rescuing several of their soldiers. This very human intervention further solidified the extant relationship between Oscar Company and the Bosnian brigades. As well, later, their rather courageous intervention was recognized with several commendations – both Bosniac and Canadian.

In-theatre training

During the tour we maintained an active in-theatre training program to maintain our combat capability, and provide a visible deterrent to the EAF. Our program included small arms training, Mortar ranges, TOW missile firing, live-fire dismounted section and platoon ranges, 76 mm cannon ranges, and mounted live-fire troop-platoon attacks. As well, we conducted a variety of low-level refresher training, and maintained interest by introducing a wealth of other initiatives such as section-level exchanges with other MND (SW) Battle Groups, an opportunity for a small group of soldiers to become trained "dog handlers" (we received the dogs from the British Contingent), and opportunities for sports teams to compete in theatre and beyond. Our tour also benefited from a visit from several Canadian entertainers and musicians, and we had a large number of visits from senior military leaders and politicians.

Visits

We had an enormous number of visitors to the Contingent – 340 in all. We were visited by our Minister of National Defence (Mr. Art Eggleton at the time), and, at Christmas, the Chief of the Defence Staff (General Maurice Baril) and the Canadian Forces chief warrant officer. Early in the tour, we were also visited by the Deputy Minister (Mr. Jim Judd) and the Vice Chief of the Defence Staff (Admiral Gary Garnett). The DCDS, Major-General Ray Crabbe (whose son, Derek, was part of our Battle Group), visited late in our tour. The Chief of the Land Staff (Lieutenant-General Bill Leach) visited briefly, and both SFOR Commanders, General Shinseki and General Meiggs, paid us visits. Both visits were memorable, given the very human touch of each of these two fine American generals. General Shinseki, impeded by a prosthesis on his right leg, visited our headquarters, and then participated in a foot patrol in Bihac at the end of which he bought the entire patrol coffee at a local café. General Meiggs arrived in time to participate in awarding SFOR medals (as well as learning Canadian drill and protocol in the process) along with our Colonel of the Regiment (Lieutenant-General (retired) Jack Vance), and our brigade commander, Colonel Matt Macdonald.

We also had the honour of a visit by the then-British Defence Minister, Lord Alistair Roberts, who later became Secretary-General of NATO. Beyond these VIP visits, we enjoyed visits by the commander and deputy commander of LFCA, and a large number of Honourary Colonels. In the main, these visits – while labour intensive, were very enjoyable, and provided frequent comic relief as we reacted and worked to prevent things from falling through the cracks. Again, since visits provided us an opportunity to showcase the achievements of our soldiers, I gave each of them a lot of personal time and consideration, hoping to leave a very positive impression in the mind of each visitor at the end of each visit.

Fatalities

We suffered two fatalities during our tour. Corporal Jim Ogilvie was killed on the highway leading to Zgon when his Cougar overturned at night, while returning from a patrol. Sapper Gilles Desmarais was killed by electrocution in a tragic construction accident in Zgon. Both of these great young men will always be remembered for their service with the 3 RCR Battle Group. Their deaths affected all members of the Battle Group, and provided considerable leadership challenges to the entire chain of command as we struggled to deal with these losses. Of all the leadership and personal challenges I faced during our tour in Bosnia-Herzegovina, none can compare to losing a soldier under one's command, in my view. The sense of loss and sadness was beyond words, and it took all members of the Battle Group considerable time to process our collective grief. Both soldiers were returned to Canada with dignity, respect, and compassion as quickly as possible.

We spent time counselling the soldiers who had been involved in the two accidents to ensure that they received the required care; however, true closure came in the form of the visits to theatre of the next-of-kin of the two deceased soldiers. Corporal Ogilvie's wife, and Sapper Desmarais' mother, each accompanied by a male family member, are among the most gracious people I have ever encountered, and their visits provided tremendous comfort and closure to the soldiers who had worked closest with the deceased members of the Battle Group. Both visits had some difficult emotional moments for all involved, but I highly recommend supporting such visits.

MAJOR TOUR ISSUES AND ACTIVITIES

Throughout our tour, we dealt with a number of significant issues, including efforts to return Displaced Persons and Refugees to their homes, judicial reform, related crises beyond the AOR, attempts to capture Persons Indicted for War Crimes (PIFWC), growing crime and black marketeering, rebuilding efforts, elections, and border conflicts. The year 1998 had been dubbed the "year of return", with the aim of either returning DPREs to their original homes, or resettling them elsewhere. In fact, despite the fact that DPRE resettlement or return remains a key ingredient of a lasting peace, it turned out to be anything but a year for DPRE returns!

The crisis in Kosovo, and the bombings of Iraq provided the occasional rise in "threat level" within the AOR, serving primarily to restrict our freedom of manoeuvre and interrupt routine operations. At one point, however, we suspected that our platoon house in Bihac was being stalked by a small team of unknown persons. That led us to use the Contingent Commander's Advisory Team (CCAT) to maintain observation on the house being used by this suspected group of possible terrorists. Eventually, the investigation petered out, and the group left the area. This little episode, aside from allowing us to gain some very useful experience at counter-surveillance operations, simply served to underline the inter-linked nature of many of the conflicts in the world, and made us much more aware of the broader threat picture. And, finally, while we had no direct dealings with PIFWCs, seizures of PIFWCs elsewhere in the AOR led to heightened alert states several times during our tour. Crime, and in particular black marketeering in forestry projects, was becoming an ever-increasing problem in Bosnia-Herzegovina.

Dealing with growing crime in the region

To assist in reducing the influence of crime in the region, and in particular stopping EAF support or involvement therein, commander MND (SW) introduced the Covert Observation Platoon (COP) concept to the region. Their first deployment took place in our AOR – after considerable planning and discussions, given the covert nature of their task, and the fact that it only peripherally concentrated on our mission of focusing on the "purely" military aspects of the Dayton Accords. This concept was relatively new to us, and I took a very personal interest in the work of this British dismounted reconnaissance platoon.

The platoon deployed in Drvar, and set up a Command Post (CP) next to the November Company CP. In the CP, the COP laid out its grid of intersecting personalities and activities, and tracked the flow of traffic passing the three covert OPs they had established on the approaches to Drvar. A special OP was established to observe the comings and goings of the senior leadership of the 1 Guards (Croatia) Brigade and its local supporters of influence. This particular OP was, unfortunately, discovered by chance by a group of children playing in the abandoned Olympic training site in the downtown area of Drvar, and had to be aborted within a few days of the operation having started. Nonetheless, the operation was a very interesting one, and provided us with lots of good lessons, and some new intelligence.

Supporting rebuilding efforts

Rebuilding had risen to one of the top spots on the agenda of the UN in Bosnia by the time of our arrival in theatre. In fact, reconstruction was well under way before we even arrived. Nonetheless, our CIMIC section played a major role in effecting the necessary coordination and liaison with international organizations to support and energize reconstruction efforts throughout our AOR. As well, through the CIMIC cell, we completed several internationally funded economic recovery and humanitarian projects, and distributed a large amount of humanitarian supplies. Not only did this contribute to better relations with a variety of international organizations; it also improved the lives of the local population, and, coincidentally, gave significant personal satisfaction to the Canadian soldiers involved. Further, many Canadian schools, communities, and volunteer groups established strong links with sub-units and platoons, which in turn provided an excellent venue for channeling humanitarian support to Bosnia.

Clearly, peace support operations involve important, but distinct contributions by civilian and military components of the mission. Throughout the tour, the Battle Group was a very active supporter of rebuilding efforts, and we therefore routinely interacted with numerous organizations engaged in rebuilding efforts such as: diplomats (including our ambassador and his staff); the IPTF (including Canadian police officers); UN agencies (UNHCR primarily); government agencies (e.g. CIDA primarily, through the Embassy); non-governmental organizations (NGOs) – mostly private volunteer organizations; and other agencies such as the OSCE.

However, we were often frustrated at the resistance we encountered when attempting to generate a more cohesive and synchronized effort amongst the many international agencies engaged in the rebuilding effort. Often, working with these organizations led to misunderstanding, conflict, and problems in general. These problems arose despite the fact that we all had essentially the same aim, that being to assist the local population. In my view, it is critical to mission success that the activities of the UN Civil Police, Election Supervisors, Human Rights Monitors, Humanitarian Aid Agencies, and similar organizations are integrated, coordinated, and synchronized with military operations – or vice versa!

Key to preventing friction is obviously the establishment of effective communications and mutual goodwill. Coordination among the components should be frequent, routine, structured, and in place at the lowest practical level, such as at the unit or sub-unit level. Accordingly, in our case, we took the initiative (almost exclusively) in organizing regular monthly meetings, at all levels. Our monthly AOR-level "Principals' Meeting" included representatives of international organizations operating in both Cantons One and Ten, the two Cantons (i.e. province-like governance structures) spanning our AOR.

The Elections

The elections were supposed to be the main focus of our tour, and did in fact figure with some prominence during our rotation. Preliminary planning for the elections included briefings at MND (SW) Headquarters by the British Division Political Advisor (POLAD), as well as formal back briefings by the COs of each Battle Group to the commander MND (SW) on national ROEs and tactics – with special emphasis on how each Battle Group would approach crowd confrontation situations. The Battle Group had already begun extensive planning efforts in the period leading up to the elections, and our plans were well advanced before the meeting at MND (SW) Headquarters. We had undertaken a considerable intelligence gathering effort, and spent considerable time visiting with the local police chiefs to see what plans they had made to support the elections. We had also spent considerable time liaising with the voting officials and the international election observers to make sure we understood their plans, and were positioned to support them effectively.

The town of Bihac, which lay within our AOR, was widely seen as being a centre of political controversy, and thus having the greatest potential for friction and violence. The main issue in Bihac was the possibility of seeing the return of a number of high profile but controversial political figures to participate in the election. For that reason, Major-General Cedric Delves considered allocating a sub-unit of the recently arrived Italian Caribinieri SFOR Multi-National "Intervention Force" to the Canadian Battle Group. I suspect, however, that he also wanted to make sure the Canadians had a source of crowd control at hand, given his subtly expressed reservations with respect to our approach to crowd confrontations. At any rate, we took the Italian Caribinieri on board, including an excellent young liaison officer who worked directly with myself. As a result, we undertook a joint training package with them to enhance our prospects for working together successfully during the upcoming elections.

Immediately prior to the start of the elections, Major-General Delves invited me back to his headquarters to provide him with a briefing on our plans for maintaining security – with special emphasis on Bihac and our planned use of the Caribinieri. I briefed him on our plans to isolate Bihac, including establishing checkpoints at all main approaches, jointly patrolling with the local police force (and monitoring their communications system), as well as placing snipers at critical junctions with a view of large rally locations. However, in the end, to our mutual amusement, it became clear that I needed to draw a very specific sketch of exactly where I would place the Caribinieri (the famous holding area "box"), and how I planned to introduce them into the downtown core of Bihac – a sketch that was ultimately faxed to SFOR Headquarters by Commander MND (SW) in response to SFOR's request for a detailed explanation of the role of the Caribinieri in the elections!

Overall, the elections unfolded without large-scale violence in spite of an interesting beginning, which saw grenade and rocket attacks kick off day one of the election in the town of Bihac. One of the two attacks were aimed at the house of a local candidate, while the other was aimed at the office of one of the local parties. These attacks gave use extra incentive to get our checkpoints into action, and soon we were confiscating weapons and various other dangerous equipments on the outskirts of Bihac. At one point during the election, in response to rising tension and public confrontations between various groups of political supporters, we increased our alert state to include wearing of helmets and body armour. The local population noted this, and responded by calming down noticeably.

Major rallies took place in the northern part of our AOR, including the towns of Bihac and Cazin, and I attended all these flashpoints personally, accompanied by our contingent commander, Colonel Natynczyk, who quietly watched the Battle Group execute its plans for maintaining security during the elections. It was entirely helpful to have the contingent commander along, more so because he was also a good friend and someone who knew the issues and the Battle Group very well, and who was adept at offering helpful advice and support while allowing the tactical commander to command the operation.

At one critical point during the elections, tensions rose to a dangerous point in Bihac, and the local police commander notified me that he was unable to contain it. In fact, two large crowds were taunting each other, and throwing various projectiles at each other in the downtown core. Given what I perceived to be an imminent threat of the two groups physically joining in a real fight with potential for significant injury and property damage, with further potential for spreading beyond the downtown core of Bihac, I asked the Caribinieri to deploy from Velika Kladusa. As they arrived, we positioned them in the vicinity of the downtown park, but with clear orders to remain within their vehicles until instructed to dismount. Eventually, I directed them to dismount and prepare to advance on the crowd on my personal order. However, as the Caribinieri dismounted, it became clear that the locals understood the skill and experience of these paramilitary policemen – the crowds immediately grew silent, and began quietly dispersing! Soon after, curious locals approached the Caribinieri to chat, and to take pictures – the Caribinieri were only too pleased to engage in conversation with the locals and to pose for pictures! The elections went otherwise as well as could be expected, given the lack of democratic experience of the region. Rallies were large and boisterous, and there were minor physical confrontations throughout the area, along with a small number of fights and beatings aimed at intimidating one group or the other. We investigated all these incidents, and followed up all local police investigations through our Canadian contingent SFOR MP Platoon and the IPTF officers.

Border problems – OP SHANNON

Primarily Croatia, but other states as well, continued to press their territorial ambitions in the face of vulnerable Bosnian security forces. In turn, we instituted a thorough patrolling program to build a database to support maintaining the integrity of the internationally recognized borders. However, ultimately, a border dispute in the Martin Brod area led us to take action to enforce an internationally recognized part of the border between Croatia and Bosnia-Herzegovina, in an operation named OP SHANNON.

At the request of the OHR (Mr. Carl Westendorp), and with the support of the Supreme Allied Commander Europe (SACEUR), Commander SFOR directed the commander MND (SW) to conduct OP SHANNON. This operation was aimed at reintroducing Bosnian authority over a disputed piece of terrain illegally controlled at the time by Croatia. Given that the disputed area was located within the Canadian AOR, Commander MND (SW) directed the 3 RCR BG to move and reposition the Croatian Border Post, and facilitate the reassertion of Bosnian authority over the Martin Brod area. We effected this very interesting operation on 22-23 December 1998, using surprise and overmatching numbers to accomplish the mission without resorting to force.

The Martin Brod area had been a source of dispute for a considerable period of time – this was certainly not a new story. However, SFOR was eager to demonstrate its resolve to the Croatians who had until then entirely ignored SFOR's direction to remove its border police from the area and return it to Bosnian control. Thus, our plan for the operation foresaw a very robust intervention requiring the complete assets of our Battle Group to rule out any outcome but the one we needed. Furthermore, given the presence of several locally employed personnel within our camps, we were aware that our success would rely heavily on operational security – the element of surprise.

First on our agenda, however, was the requirement to gain national support for this "aggressive" operation that would see elements of the CF possibly violating sovereign Croatian territory. Colonel Natynczyk paid a visit to the British Commander of MND (SW), and quickly thereafter lent his support to the operation. The day before the operation, I personally received a telephone call from the CDS to confirm his support for the operation, and to receive his personal wishes for mission success.

The task itself was a complex one – not only due to the sensitive nature of the tactical and political situation, but also because of the difficult terrain we were facing. The village of Martin Brod lay at foot of some impressive cliffs, at the bottom of the Una River valley, in a rather narrow re-entrant, and connected to the Bosnian side of the Una River by an expedient military metal bridge installed in place of an earlier bridge that had been destroyed during the earlier fighting. The few approaches to the area were narrow winding roads following the flow of the river, or carved precariously into the sides of the mountains surrounding the valley on both sides of the Una River. We spent considerable time studying the situation, and becoming familiar with the area – without tipping our hand in the process.

I ultimately planned to have the Reconnaissance Squadron, with a platoon from Parachute Company (commanded by Warrant Officer Mark Godfrey) under command, lead the way across the bridge spanning the Una River. After considerable reflection and discussions, I had elected to have this platoon effect entry into the border post buildings, rather than the Joint Task Force (JTF) 2 section operating in our area. Instead, the CCAT came along as a very useful Combat Camera Team, who recorded our actions carefully to make sure we could deal with later allegations of impropriety should we need to. This initial assault group ("A" Squadron Combat Team) would be followed immediately by a large group of engineers whose tasks included dealing with any explosives or mine threats, as well as removing any obstacles, displacing the Croatian border installations, removing all fortifications, and then, further West, establishing a Bosnian border post.

Parachute Company (minus) became responsible for the eastern side of the Una River, including notifying the local residents of the change in government effected overnight, following the operation. The liaison and CIMIC section officers were assigned key specific government and police personnel to notify during the night, following the successful completion of the initial phase of the operation – with a view to having these officials move to the area to assume responsibility for governing it as we returned to routine operations.

November Company became the Battle Group Reserve with orders to seal off Drvar and 1 Guards (Croatia) Brigade. Major Walker was directed to visit the Guards Brigade Commander during the night to notify him of the change in circumstances at Martin Brod to prevent any misunderstandings and possible friction. Oscar Company provided security at all border crossings to the North of the objective area, along with a visible presence in several key towns, villages, and junctions. The Anti-Armour Platoon provided over-watch at key border crossings on the flanks of the operation, given the proximity of the Croatian Army's ability to intervene in Brigade strength from the Knin area (about 60 kilometres from Martin Brod), and the Mortar Platoon was established immediately North of, and within range of the Martin Brod area. I retained some anti-armour assets on the high ground East of the objective area to provide me with long-range observation (including thermal sights) and engagement options. I brought along an electronic warfare detachment, along with interpreters, to monitor radio traffic in the area – this paid dividends as the operation unfolded.

CHAPTER 3

Camp security was provided by troops of the NSE, an option my good friend and colleague, Lieutenant-Colonel Ryan Jestin, the commander of the NSE, had practiced in Canada during work-up training. The NSE troops ultimately moved into the camps at Coralici and Zgon without fanfare, and carried out this critical function with great professionalism. MND (SW) sent a communications detachment with my headquarters, and we were expecting overflights of reconnaissance aircraft and even an attack helicopter near first light! However, the commander MND (SW) had refused my request for a battery of artillery in "direct support" in the event of a Croatian intervention in force from Knin. I also had Canadian helicopters in support, for command and liaison purposes, as well as for casualty evacuation purposes.

We had prepared detailed orders for the operation, and issued these under very elaborate operational security conditions. Although I had desperately wanted to rehearse the operation, security precluded us from doing so. Instead, the day before the operation, we had all section, crew, and detachment commanders and above meet at our platoon house near Bos Petrovac where we could conduct a detailed back brief without the presence of locally employed personnel or other potential security risks. During the back brief, we used a model to help the process, and had each key commander and leader walk the entire group through his part in the mission in detail, in chronological sequence, starting with "A" Squadron. I found this process highly useful, and I emerged very confident from this gathering. The RSM and I spent the night with November Company in Drvar.

The first attempt at executing this operation was aborted shortly after the various elements of the Battle Group had deployed from their camps. The order to abort came from commander MND (SW), now Major-General Rettie Watt, who felt operational security might have been compromised. This was a significant disappointment – but served to remind us of the requirement to practice some basic security precautions such as shredding all paper correspondence, watching our conversations on the telephone and the radio, and generally paying closer attention to our surroundings (i.e. locally employed personnel). However, shortly after, we were told to prepare to execute the operation again – this time on short notice. We began the operation at midnight on 22 December, with "A" Squadron leading across the Una River at 0200 hours on the morning of 23 December.

The operation proceeded as planned, with no resistance encountered from the Croatian border police who exited the border post with their hands held high immediately after the building was surrounded by the Cougar Squadron, who used their high-powered search lights to great effect. The platoon from Parachute Company took the border police into custody, disconnected all communications means, and immediately took down the Croatian flag, and transported the police officers and all their belongings to the first town in Croatia where they were released to contact their superiors. "A" Squadron, under Major Mike Nixon, immediately exploited to the point of territory we had been directed to secure for the Bosnian border post. Immediately behind him came the vehicles of 24 Field Squadron, loaded with sandbags, border huts, gates, flagpoles, etc., for use in the construction of a formal border crossing point several kilometers West of the Una River.

The instant "A" Squadron reported being in position, I directed the liaison and information aspects of the operation to commence. The Engineers worked superbly in the dark to establish a proper defended locality in the vicinity of the new crossing point, and Warrant Officer Godfrey's platoon returned to help "A" Squadron provide security in the area near the new border crossing. Near first light, we intercepted radio communications immediately West of the new border. These communications originated from border security troops on the Croatian side of the border, and used the call signs "Tiger (One, Two, etc.)". We spotted several paramilitary vehicles arriving in the area – but all remained well away from the new border location.

Soon, our Thermal Imagery revealed that several two-member groups of dismounted personnel were deploying across the front of "A" Squadron Combat Team. Given that these dismounted personnel appeared to be carrying long rifles, I ordered the "A" Squadron team to "button up" and remain under cover and be prepared to respond. Our radio intercept capabilities proved highly useful – we determined immediately that we had achieved complete surprise, and that the Croatians were less than pleased at the turn of events!

Shortly after first light, the RSM and I deployed to conduct an over-flight of the area using one of our allocated Griffon helicopters. The very talented pilots landed the Griffon at a tiny road intersection, and we flew over the area aided by forward looking intra-red (FLIR) television screens to confirm the locations of the Croatian border forces. Our intent was simply to demonstrate that we had complete coverage of the area, and that we were aware of their deployment. Shortly afterwards, a fixed-wing reconnaissance aircraft flew over the area, followed by a slow-moving SFOR attack helicopter. This prompted the Croatian border police to pull back from the area, and allowed us to introduce the Bosnian officials to their new area of responsibility before noon on 23 December 1998.

Over the next 24 hours we experienced several small incidents which caused us concern and led us to raise our alert status, but eventually we were able to turn over the area to Bosnian border police completely, and withdraw a very cold and tired Reconnaissance Squadron from the Martin Brod area early on Christmas Day, 1998. Timing was perfect – the CDS had arrived to help us celebrate Christmas, so we were able to turn our attention to his visit. Subsequent to the successful conclusion of this operation, we encountered delays and harassment at the crossing points – including the main administrative crossing area – with Croatia. After a period of time, however, operations returned to normal.

REDEPLOYMENT

At about the halfway point in our tour, we began preparations for redeployment to Canada. This was about the right time for us to begin this process of refocusing and restructuring. Human nature dictates that individuals need to have a new horizon presented at regular intervals to help maintain focus and interest, so we began our deliberations for restructuring of the unit well before redeployment, aligned with newly assigned Canada-based tasks for our return. Our handover and formal "Transfer of Authority" (TOA) to the 2nd Battalion, The Royal Canadian Regiment (2 RCR) took place on 30 January 1999.

This process unfolded very smoothly, given that we were all very well acquainted. The return trip proceeded uneventfully, and the Arrival Assistance Group was superbly organized to process us quickly and return us to our waiting families. The leave plan had been well orchestrated in advance, and the chain of command had been very generous. We had worked especially hard to find a way to have the folks from outside Petawawa returned as quickly as administratively possible to their homes as well, but this proved challenging and created some dissatisfaction among these individuals who had served so well alongside the rest of us.

We had completed personnel assessments on all Battle Group personnel – deployment Personnel Evaluation Reports (PERs) – well before we returned, so the formal recognition of the non-Petawawa personnel had been completed (later on, we also completed annual PERs on the members of 3 RCR, while parent units did the same for ex-Battle Group personnel!). Shortly after returning to garrison routine, we spent some time thanking local community groups and organizations for their support to the operation – both for their support to the deployed component, and for their support to families and Rear Party personnel. As well, shortly after redeployment we had a unit party to celebrate our tour – in this case, we tied it into Regimental celebrations of the Battle of Paardeberg, and held it late in February.

CONCLUDING REMARKS AND OBSERVATIONS

Cataloguing our achievements

In reviewing our accomplishments with the benefit of 20/20 hindsight, my sense is that key accomplishments include at least the following aspects of the operation:

- A successful "Hearts and Minds" campaign
- OP SHANNON
- An effective in-theatre Training program
- A record-setting Civil Development funding program
- Marked improvements in cooperation with, and professionalization of, the "Entity Armed Forces"

Beyond the above list of key accomplishments, I also feel a brief discussion of a series of issues is pertinent to this overall summary of our experience in Bosnia-Herzegovina in 1998-1999. They are in no particular order of significance, but all remain strongly etched in my mind as being important "lessons".

The importance of having unimpeded freedom of action and movement

We enjoyed complete and unimpeded freedom of action and movement. The Status Of Forces Agreement (SOFA) governed this aspect of our mission, and guaranteed that we could execute our mission successfully. This aspect of our mission differed significantly from my earlier experience in UN missions where the EAF/belligerents seemed to hold the balance of power. Clearly, my view is that this unrestricted deployment is critical to mission success – regardless of mission, and should be a precondition to any Canadian deployment in support of stability operations. On the other hand, I should note that other agencies and NGOs were routinely harassed at crossing points and in other locales.

We never negotiated with any EAF, or conducted mediations between EAF. The NATO-led SFOR simply did not negotiate – we worked in a zero-tolerance, enforcement-centric command climate in MND (SW). Having said that, I should emphasize that we always maintained a productive dialogue to preclude miscommunications leading to unnecessary confrontation.

Force Protection

The issue of force protection played a constant, and appropriate, role in our planning. While not being a proponent of intervention forces who must rely on the good will of the EAF/local militia, I am certainly a proponent of the concept of using force only as a last resort to prevent the unnecessary risk of injury or death for our soldiers. Careful planning can reduce the risk of having to resort to force – while not preventing or reducing mission success, and I believe this is well illustrated in our execution of Operation SHANNON, which had considerable potential for things going astray.

Dealing with the Entity Armed Forces

We took every opportunity to be courteous to, and learn about, the entity armed forces. At the same time, we worked hard to discharge our duties smartly and with the utmost professionalism, including live-fire exercises and demonstrations – all of which the EAF commanders noted carefully. This approach was not just professionally rewarding, but it helped to create an atmosphere of mutual respect that allowed us to progress several small initiatives relatively quickly, including efforts to "professionalize" the EAF in order to render them less prone to sudden outbursts of violence and disloyalty to their rightfully elected political masters.

We were very conscious of the need to maintain our reputation of impartiality throughout. If we played soccer against one EAF team, we would make sure to replicate the activity with a different EAF team, from the "other side", shortly thereafter. Similarly, if we presented a gift or any token to one side, or attended a cultural event hosted by any EAF, we would immediately make sure we reciprocated or participated in a similar capacity with the other EAF. Again, the chain of command needs to make sure this concept is well understood at every level of command within the unit.

Credibility is another critical component of mission success. Credibility is very much tied to the issue of mutual respect, and underscores the need for units to arrive trained for general-purpose combat operations, precisely as we had done. We left no question in the minds of the EAF with regards to our ability to discharge our responsibilities – we were disciplined, thorough, and highly effective, both on and off duty. Our alcohol policy was part and parcel of this important dimension of operations in Bosnia, as was our policy that precluded folks from leaving the camp while off duty ("walking out" policy).

While we were very deliberate in our operations, and generally conducted our routine operations in the most transparent manner possible, we never acted in a manner that would jeopardize our own security. This approach is critical to preventing unintended consequences arising from poorly communicated intentions – and, again, it is closely linked with the importance of maintaining impartiality, credibility, and mutual respect. Certainly, in rare instances we proceeded "covertly", including OP SHANNON, the Covert Observation Platoon tasks near Drvar, and our use of electronic warfare (EW) assets to monitor activities in the area around Bihac, where we felt the platoon house was being "cased". As well, we conducted a few no-notice inspections of selected cantonment sites to keep the EAF commanders on their toes!

Unity of Effort in Operations

Generally, we enjoyed a high level of unity of command in our tour. We received clear direction, and had a good level of interaction with our tactical commander, Commander MND (SW). However, beyond that – and, in particular in the case of the IPTF, I felt the greater SFOR mission lacked a sense of "unity of command". The IPTF often proceeded unilaterally, and thus failed to benefit from the synergy and total effect a united front could produce. In my view, their work fell far short of its potential for success.

The challenges of out-of-area operations

Although the Canadian contribution to SFOR was restricted to operating within its own AOR, we were able to support some activities beyond the routine tasks after passing these through the national chain of command – OP SHANNON being an example thereof. Having said that, I feel that we might have been more effective, and valued more by our Allies, had we had greater flexibility to participate in, or mount, out-of-area operations. My sense is that tactical commanders of the future will need, and get, greater flexibility in future theatres of operation to prosecute operations without an overly strict limitation on their ability to operate beyond the lines drawn on their map, representing their "AOR".

Administrative and Medical Support

I felt we were well served by the NSE. The NSE, an element of the Canadian contingent SFOR, provided support beyond the levels provided by NATO, by virtue of Canadian standards of health and safety. To my mind, this was a real bonus! Our soldiers were superbly cared for in terms of food and other logistics and amenities. Similarly, we were served in a manner second to none by our medical system – the presence of this team of medical experts provided a critical sense of soldier confidence within the Battle Group. This was especially obvious during the two tragic fatalities, but we also had other medical emergencies (e.g. serious heart problems, serious digestive system problems) that benefited from this incredible medical resource.

Communications

On the other hand, I felt our communications equipment was far from effective. Our technicians are superb, and can move mountains. However, our communications remained generally unreliable in this difficult AOR, which is surprising given the progress of technology we see about us daily. In particular, I recall a surprise cantonment site inspection we conducted where our communications plan involved placing a string of AVGPs acting as "RRBs" and "step-up" communications nodes between the controlling command headquarters and the platoon undertaking the inspection. My view is that we need to correct this problem immediately, given the constant negative impact it has on all tactical operations.

Information Operations

Our public affairs and public relations efforts were frequently sources of conflict and friction. I would strongly recommend that commanders take a very firm and active hand in all public relations and public affairs issues, given that the potential for disaster is significant if these activities are left to unfold on their own. I remember very well several successful media visits, but I remember equally well a less-than-successful visit from a female reporter from a Toronto newspaper to Parachute Company at Zgon, and some friction arising from an otherwise enjoyable visit from Canadian researcher, Dr. Donna Winslow. And, I am afraid that my own weekly, taped (on Sunday mornings) operations summaries were not as informative for the families left in Canada as I had hoped!

Our fledgling public information campaign – conducted in close cooperation with the larger SFOR public information campaign, on the other hand, showed considerable promise in my view. Notwithstanding a few, early, misguided efforts at jazzing up the campaign (playing "I used to love her, but I had to kill her" by Guns'n Roses as a metaphor for Bosnia-Herzegovina on the local SFOR-supported radio station), Captain Phil Millar did a remarkable job at kick-starting this relatively new tool in our tactical tool box, designed to influence public sentiment, which is a promising tool in stability and peace support operations.

Quality of Life

Overall, I felt our quality of life was excellent – given the context within which I make that assessment. Our rations were second to none, as was our medical support system. Our accommodations were generally pretty good – although they were not universally as comfortable as we might have wanted them to be. Our Welfare and Home Leave Travel Assistance (HLTA) program was outstanding, in my view – noting that the key to success was to plan early, and insist on folks taking regular breaks (i.e. 72-hour passes to various "R&R centres"). My only regret, in fact, in terms of human resource issues, was my failure to sufficiently recognize the exceptional performances of the many individuals in the Battle Group, in terms of honours and awards, and commendations.

Maintaining Discipline and Good Order

The establishment and maintenance of discipline within the Battle Group was relatively straightforward – our soldiers, non-commissioned officers, and officers were high-calibre, well motivated, and highly dedicated individuals. We did have a handful of minor issues to deal with during the tour – a very reasonable expectation over a six-month period for 883 soldiers on tour. Interestingly enough, most of the incidents transpired at the outset of our tour, with a few taking place towards the end of the tour. Alcohol was a not a real issue – the two-beer per day policy was a reasonable compromise; however, I did find the insulting way in which we had to implement it to be somewhat demeaning. But, as I could not think of a more "certain" way to ensure compliance, I had to work within the established procedures.

We experienced our first disciplinary incident upon arrival – immediately after we had taken command of the AOR. The outgoing liaison officer from our predecessor 1 RCR Battle Group was returned from Banja Luka by the Chief of Staff of MND (SW) as a result of the complaints made by a female officer who had witnessed the Canadian (male) officer visiting her (female) room-mate in the trailer accommodations, contrary to regulations precluding male visits to female quarters. This case remained with us for almost the entire duration of the tour. Part of the problem lay in the fact that the contingent legal officer recommended against the laying of charges from the outset. This was extremely frustrating – to myself, as well as to the contingent commander who shared my perspective that this disciplinary and ethical transgression deserved the laying of charges.

Ultimately, charges were laid, and the officer was provided with the choice of court-martial, which he subsequently accepted. Aside from leaving me with a feeling of being poorly served by the military justice system and our legal advisors, this incident created significant doubts at the start of our tour that the chain of command would not support the efforts of the non-commissioned officers to maintain discipline. Subsequently, I decided to hold a special meeting of the "War Cabinet" to address the concerns of the Company Sergeants-Major with respect to the difficulty inherent in maintaining discipline in the traditional way. This meeting went a long way towards shoring up confidence at the "tactical" level that we had the tools necessary to maintain good order and discipline in the Battle Group during the tour. I later communicated my concerns to the Judge Advocate General, and received a candid, written reply from him in turn.

We also had a case of "negligent discharge" upon arrival. A Senior NCO on duty at Camp Coralici fired a negligent discharge into the "unloading bay" container during the night. Unfortunately, there was no way around laying charges, but the process was a painful one for myself, as well as this very earnest individual who had reported his own error to the duty officer immediately after the incident! Luckily, a tiny punishment met the need to take action.

We had spent a fair amount of time selecting our Rest and Recreation (R&R) location, and even longer convincing the hotel owners to let us become the very first guests at their brand-new, state-of-the-art hotel. It was a very impressive hotel, and it offered our troops a first-class location for unwinding on the Adriatic coast. Regrettably, on the very first R&R trip conducted by our contingent, a soldier was arrested by Croatian police for having tossed drinking glasses from his ninth-floor balcony onto the cars parked below! Our Military Police returned the soldier to the AOR immediately, and I had him returned to Canada as quickly as administratively possible, and replaced. Of note is the fact that this particular individual was a border line selection for the Battle Group, having had some previous disciplinary issues on his file.

The most challenging allegations I had to deal with on the tour were two separate incidents dealing with the use of "uncleared" routes. In one incident, the TOW under armour (TUA) Detachment Commander was reported to have taken an "uncleared" track, after having become temporarily lost in a difficult piece of terrain. In a second incident, an Engineer Detachment Commander was reported to have made use of an unapproved route – raising questions about our definition and understanding of what constituted an approved route for travel. In both cases, I felt there was sufficient ambiguity to warrant findings of "not guilty"; however, both cases took considerable time to hear, and were very challenging. In both cases, we learned more about our policies dealing with the "proven route trace"!

Later in the tour, I had to deal with a significant theft of beer from the Zgon all ranks mess. The Military Police conducted a thorough investigation, but unraveling the case proved difficult. In the end, we managed to put an end to the thefts (conducted by effecting illegal entry through the roof of a wire cage!), and lay some charges in the process – although the identity of the culprits surprised us. This incident was followed by allegations that two senior NCOs had exceeded the limit on alcoholic beverage consumption on one occasion; in the end, an absence of evidence precluded the laying of charges. Not so with another senior NCO, who while on patrol decided to have a drink with some local members of the community to say farewell. He was found guilty, reduced in rank, and returned to Canada, having significantly damaged our credibility, and put the members of his section (and their weapons and equipment) at risk.

SUMMING UP

Looking back, I firmly believe that we had a very successful tour. I accredit that to excellent leadership at all levels, but in particular the junior leadership who provided the critical direct leadership in this "section commander's operation". In many ways, I feel that observation supports my view that we had succeeded in our efforts at personnel selection and training. That process – including personal interviews, was time consuming, but well worth it. In the end, I found that you need to remain ruthlessly selective.

In the process of participating in the Board of Inquiry into the death of Sapper Desmarais, I did learn that I had perhaps taken safety for granted – most of us probably do take "routine" safety for granted. My sense is that I should have been more proactive in stressing even the very minute and minor aspects of routine safety, especially as the tour proceeded towards the halfway point and we all became naturally a bit more complacent in anything but operational safety and security.

The greatest leadership challenge I faced, as I have stated earlier is that of dealing with fatalities or serious injuries amongst members of the Battle Group. I do not think we are well "trained" to handle casualties beyond the technical processing thereof. In terms of how we communicate these incidents to the other members of our team and deal with the strong emotions released in the wake of such a tragedy, however, I feel we need to become better prepared. I was immensely impressed with the strength and grace of the next-of-kin of both young soldiers who died; their families were a source of inspiration to the rest of us during their visit to our Battle Group, although I personally felt apprehensive about the visit prior to their arrival. My only advice to our leaders is to become personally engaged in the process from the start, and make sure to personally connect with the next-of-kin.

The "Command Team" consisting of the Regimental Sergeant-Major and the Commanding Officer must be extremely visible to be effective. I believe that Chief Warrant Officer Frank Grattan and I took every opportunity to spend time with individual members of the Battle Group, and in the process came to know most of them fairly well. This is a very enjoyable aspect of command, and we tried to spend one day and night at each camp every week, and our goal was to conduct at least one patrol with every section, crew, and detachment in the Battle Group during the tour. In fact, we managed more than that, and had a lot of fun doing it. I also believe that being so directly available at all times allowed us better insight into the goals and concerns of all individual members of the Battle Group. One thing that struck me was the requirement to reinforce the successful nature of our mission within our soldiers. On that topic, I simply emphasized that as for measuring success there really are no definitive characteristics of a successful end-state for this type of operation. Commanders must judge the success of their operations according to their specific mission. Armed forces alone cannot solve the political problems underlying this conflict. That is the job of diplomats, politicians, economic specialists, etc., and there is no guarantee that they will succeed. However, the indications of (long-term) success are that negotiations are in progress, while fighting has stopped or been significantly reduced. Nonetheless, if peace support operations are to have any hope for success, the professional execution of their military component is a necessary pre-condition. In our case, we approached this more simply, basing our measurement of success to some degree on the quality of the handover to our successors. We certainly felt we had had a highly successful tour, thanks to the enduring high quality and dedication of Canadian soldiers, and the truly sound leadership at every level of command within the Battle Group.

CHAPTER 4

OPERATION PALLADIUM ROTO 13, SEPTEMBER 2003-APRIL 2004

Colonel Dean Milner

I was the Commanding Officer for the Royal Canadian Dragoon (RCD) Battle Group (BG) for Operation (OP) PALLADIUM, Rotation 13, in Bosnia-Herzegovina from September 2003 to April 2004. Prior to this tour, I had participated on two earlier missions. The first was OP CAVALIER as part of the United Nations Protection Force (UNPROFOR) from September 1994 to May 1995. I was the Officer Commanding (OC) "A" Squadron – Reconnaissance Squadron based out of Visoko. The other mission was as part of UNMEE from March to September 2001, where I was the Chief Operations Officer for the 220 man Observer force based out of Asmara. Needless to say, these missions assisted in preparing me for my current task as the Commanding Officer (CO) of the Canadian BG for OP PALLADIUM.

In addition, I was fortunate in that I had the opportunity to conduct a reconnaissance prior to the commencement of my pre-deployment training. This was critical and, as a result, I was able to focus my training. On the other hand, as with most missions, it is difficult to acquire a full understanding of the threat and the intelligence situation prior to the deployment. As a result, the first month of most missions is concentrated on confirming the situation and fully understanding the posture of local forces.

During the mission, my main responsibility was commanding the BG during its everyday tasks, as well as during specific operations. The tempo was busy and the BG conducted over 30 operations, although mainly at the Squadron (Sqn)/Company Group (Coy Gp) level. Information Operations were key as this was the primary means of directing future tasks. As a result, I formed a strong and capable BG headquarters (HQ) of about a 100 personnel that focused and synchronized all tasks and operations, which were all generated from a refined weekly information operations campaign plan.

There were a number of challenges during the deployment, but I must admit that my biggest fear prior to the mission was complacency setting in with the troops during our tour. In the end, this was never a problem. If there was down time, I generated competitions, training, and plenty of sports and physical training (PT) activities.

One of my biggest challenges, however, was that my company was composed entirely of reservists. This sub-unit was assembled just prior to pre-deployment training and consisted of soldiers from 24 different units across Ontario. They required a large amount of supervision and my requests to augment the company with my Regular Force Senior Non-Commissioned Officers (Sr NCOs) and officers were all denied. As a result, I devoted the majority of my time with this company to attempting to ensure success, resulting in less time being focused with my other sub-units. All went well in the end, but it was a major concern for both my Regimental Sergeant Major (RSM) and myself throughout the deployment. For instance, it was a difficult decision to locate squadrons and company throughout the area of operations (AO), and after having completed a mission analysis, I deployed my reserve company to an area, which later ended up being a particularly vital one. They performed well, but at the same time, other key BG assets may have been more successful in that area of responsibility.

Our mission statement mandated that "the RCD BG will secure the Area of Operations as a safe and secure environment." Our major role was deterrence, but as this mission evolved and stability and normality progressed, the focus of the mission centred on improving the rule of law and developing defence and security capability. This primarily involved conducting surveillance and information operations (IO) to disrupt criminal and terrorist activity. Continued training and improvement of the local defence forces was also one of our major undertakings.

Harvest operations (i.e. removing weapons and explosives from the environment) were also conducted on an ongoing basis throughout the tour. Quite often, as a result of successful info ops, large weapons caches were found. Overall, as stated earlier, the BG conducted over 30 operations either at the Sqn/Coy level, or BG and Brigade (Bde) level.

The nature of the operational environment was generally stable, which is not surprising given the emphasis of the NATO Security Force (SFOR) mission, was to sustain a safe and secure environment. However, the economy was weak and fledgling, and progressing at a very slow rate. There still remained large-scale black market activities and, due to the country's porous borders, there was potential for terrorist training and activities. In addition, the host nation defence forces were being reduced to the point of almost insignificance and at the time the three forces in Bosnia-Herzegovina trained separately and were reluctant to conduct combined training initiatives. It was my assessment that Bosnia-Herzegovina could still potentially spiral back into conflict without specific international economic help and direction over the coming years.

In any case, there were a number of specific challenges to the mission. For example, as the mission had evolved, a relatively safe and secure environment had been established. As a result, SFOR gradually shifted its operational focus towards policing-style tasks. This was a justifiable emphasis based on the SFOR mandate/mission. It was also an excellent opportunity to focus on information operations that saw the BG working closely with the brigade staff and also the international community and local police forces. The obvious challenge was that this was generally new territory for my troops, but at the same time it forced the BG to concentrate on information gathering and surveillance operations, as well as the use of our Intelligence Surveillance Targeting Acquisition Reconnaissance (ISTAR) capabilities.

Our BG resources combined with brigade and SFOR resources enabled us to conduct some outstanding and successful operations. During some of these missions we came across significant caches of weapons and links to terrorist and illegal activities. Of note, SFOR, based on the Canadian and Multi-National Brigade North West (MNB NW) initiative, developed Monitoring, Observation, Surveillance Teams (MOST), which are now called Liaison Observation Teams (LOTs). These teams live in villages and were extremely successful. They worked closely with local authorities and the population in targeting information for the new task force that was stood up to conduct operations. In the end, this concept enabled SFOR to draw down its forces and subsequently deploy small teams into critical areas to measure stability and progress.

First and foremost, Canada has contributed immensely to peace and security in Bosnia-Herzegovina over the past 12-14 years. The result was evident. The situation in the country had improved ten fold during that timeframe. Unfortunately for me as a BG commander it was evident from the start of the deployment that a great deal more needed to be done. Unfortunately, there is still a deep hatred that persists amongst the people. This continues to impede the move towards a stable and normal democratic country. Although our impact was significant, I believe, that the emphasis should be more focused on economic development, as the country is experiencing no real privatization or new economic growth.

In that environment, it is not surprising that the black market is one of the most ongoing growth industries. Although it is important to note that NATO was a very efficient organization and it is obvious that it has had a very positive and everlasting impact on the peace in Bosnia, more is required. Based on Bosnia-Herzegovina's ongoing problems, it is necessary that the international community remain engaged and more effort is required to transform and evolve the country's police, judicial and other institutions.

Nonetheless, the BG worked well within the multinational brigade context. Our battle rhythm or operational daily routine was effective, so much so, that the Brigade adopted a similar battle rhythm. The BG conducted a large number of successful operations and at the same time we, along with the National Control Element (NCE) and the National Support Element (NSE), were kept busy with the draw sizing of the mission. After this rotation, the mission was scaled back from a BG sized contingent to a sqn/coy sized entity. This entailed detailed and busy planning and coordination between all elements, conducted concurrently with a high operations tempo. At one point, particularly during the later portion of the mission, it actually became somewhat problematic. It was no longer clear what the priority of the task was: operations or draw down? This caused some tension between the NCE and ourselves, but it was generally mitigated by the NCE and NSE commanders and myself.

Due to the nature of the mission and how it has improved through the years, the majority of our tasks revolved around information and intelligence gathering. Some of our pre-deployment training encompassed the type of training which helped the BG work with the multi-national brigade. However, I believe more focus should have been added to our information gathering capacity, as well as how the BG could have better worked jointly with other assets within SFOR. As usual, you quickly adapt to all situations, however sometimes this does not allow you enough time to achieve all that you would wish to.

In the end, I would emphasize the point that our troops are extremely well trained to conduct missions of this nature and they readily adapt to a wide variety of challenges. More importantly, our soldiers have learned to communicate and operate positively with the locals to the point that a genuine trust and respect has been developed in Bosnia-Herzegovina over the years. Our soldiers have been fortunate to learn the lessons that have been produced from previous missions.

For my peers: it is matter of preparing for the specific mission that you will embark on. They are all different, but the building blocks are there and it is a matter of putting together a well-refined and motivated team that is ready to meet all challenges. It is vitally important to get your personnel together as early as possible to build a strong team. Our challenges in the future will be focused on learning to operate with the new technologies that are being introduced to the army and how we maximize their potential. Examples of this are the new ISTAR capabilities that are being introduced to the Army.

For my superiors: I believe that our biggest challenge is technology and change. They are both essential but we have to ensure that we continue to incorporate them advantageously over the upcoming years. Our strength remains our soldiers, and although new technology will help us improve our ability to find and destroy the enemy, it must be well integrated. Technology cannot replace cohesion. There is no substitute for taking the time to put all the essential elements together to train as part of a BG for its future deployment. Unfortunately, it seems there is a growing tendency for elements to train with their technology separately from the remainder of the team.

In the end, the bottom line is that I established a strong team with a high *esprit de corps* within the BG and this helped me solve any of the challenges or difficulties that arose during the tour.

CHAPTER 5

COMMAND IN COMBAT

Brigadier-General D.A. Davies

Military careers are filled with hard work, professional development and training, mental and physical challenges, good times and tough times, and at least a certain amount of "right place at the right time." While most of us will be called upon to deploy on operations during some time in our career, few will be actually called upon to engage in combat operations, despite the lifetime dedicated to being ready to do that very thing. Those who are will find the experience brings a curious mix of emotions, from excitement to fear and fatigue. Fundamentally, the experience brings about a sense of validation of a career; it helps answer the question of "am I good enough," or perhaps more importantly, "did I dedicate my working life to doing something of value?"

In the spring of 1999, I had the honour and good fortune to find myself commanding Task Force Aviano (TFA) during a period of rising international tensions that ultimately culminated in our being ordered into combat against the Former Republic of Yugoslavia. The ensuing weeks saw my force grow to 18 CF-18 fighter aircraft and 350 personnel heavily engaged in combat operations, conducting the most intense aerial bombing campaign carried out by the Canadian Air Force since the Second World War. The fact that so many missions were flown without loss or accident, and that so many bombs were accurately dropped without significant incident of collateral damage is a source of immense pride to me, and to all Canadians.

What follows is not an academic treatise on leadership or command, replete with footnotes or endnotes. Rather, it is a compendium of thoughts, observations, and advice based on my experiences. It is my hope that a portion of this will be of value to some of the readers; I accept that all of it will not necessarily be of value to any one individual. Situations change, the challenges each of us faces are unique to our particular experience; despite this, some fundamentals are worthy of study.

CHAPTER 5

If I had to summarize my experience in one paragraph, I would say that my experience showed me that a military career in the Canadian Forces (CF) does prepare you extremely well for the challenges that you face on operations. I would add that this is true for the personnel at all levels in your force. The most frequent comments that I heard from my folks, from all levels, showed how much they felt that combat operations felt identical to the training exercises. As the commander, I initially found the deluge of complex issues that I had to address somewhat daunting. To my surprise, no matter what the issue, I knew how to get the team going to overcome the challenge and move on. Somehow, my career to that point had exposed me to so much that I could always provide the guidance to my equally skilled folks that would help them identify, and deliver, the right solutions.

THE BUCK STOPS HERE

While it is self-evident that the commander is ultimately responsible for his force, nothing quite drives this home as much as being the overall force commander, operating on the other side of the world. Our normal existence leads us to feel that we always have a superior to whom we can refer the really tough problems, or at least to whom we can turn for advice. When you command in operations, you may well report to a superior that is on the other side of the world, multiple time-zones away, immersed in the frenzied life of headquarters where you represent but one demand on his time. While you do have top priority for his attention, you will rapidly realize that most issues that you face cannot be decided, nor even advised, by him. He simply cannot understand the nuances or specifics of the situations that you face.

It is a rather daunting realization that you are the end of the chain, and must not be found wanting when your people turn to you. Where you have little recourse for advice and direction, they will, of necessity, turn to you for help, direction, and guidance when they face the most complex of issues, the most difficult of problems.

At times such as these, I think that it is critical to keep in mind that you do have significant talents and solid experience, and were specifically selected to lead. It is your duty to identify the solutions and issue the clear direction to your team that will get them over the hurdle and moving forwards towards mission success. While this initially feels as it must for a tightrope walker on the first crossing without a net, as you settle in and gain additional self-confidence, this will be one of the most satisfying aspects of your command. The successes of your folks are a combination of their talents and energy, and your ability to guide them. The failures are a measure of your shortfall in foresight, planning, or communication.

EVERYONE IS CRITICALLY IMPORTANT TO SUCCESS

In our organization, particularly during our peacetime preparations and training, we have a number of systems that differentiate between people. These are highly visible in some cases, such as rank and uniform, and less so in others, such as job specialty. Officers have their own messes, fighter pilots wear different uniforms, Military Police have special status, the list of differences is endless. While it may sound like motherhood, and many in leadership roles do not initially fully believe it, but every individual is absolutely key to success.

The missions cannot succeed if the technicians have not prepared the aircraft, if the armourers have not correctly prepared the munitions, if the supply tech has not found the solution to getting the required parts out of the distant supply system and into the hands of the technicians, or if the administrative team has not prepared the necessary processes to ensure that the force is properly supported. Fundamentally, you will need every individual on your team to give his or her utmost. It is critical to the team that you understand this from the outset, and that you exude this sense in all that you do and say.

As a corollary, you need to apply your considerable talents and power to help those in your team who will need assistance with either professional, or more importantly, personal issues. They will represent only a very small percentage of your force, and the temptation to focus your energies on what appears to be larger and more pressing operational issues will be great; you must resist this at all costs. You owe it to all of your folks to take the best of care of them and to be there when they need you; they have entrusted their lives into your hands. Of equal importance, the majority of your troops will see how you treat those in need, and will judge you by your actions. Even though they may never be the ones who need you, they will drive themselves all the harder if they are secure in the knowledge that you will be there for them, if they are ever in need.

I was asked about the impact of cultural or gender influences on the mission. I had to reply that I had not encountered any issues in either of these areas. In fact, when further questioned, I had to carefully think to identify who was male or female in my force. I had quite simply not thought of them in that matrix. They were all critical team players, thoroughly professional, and highly motivated.

SAFETY

While we have all read commander's visions on safety in the past, and we have all heard of the high losses in combat due to accidents, we are fundamentally all driven to succeed in our missions. There is a very fine line where this desire to succeed can motivate individuals to push too hard, cut corners, and otherwise take on levels of risk that are inappropriate to the overall success of your mission. As the commander, you play a critical role in ensuring that the right safety attitude is pervasive in your command. This goes way beyond publishing a cleverly worded safety vision on posters for distribution. Few will "waste" the time reading such posters, when the mission is calling. You must personally and convincingly communicate your expectations.

I believe that a proper approach to safety is fundamental to your success in combat. In my case, I had examined this aspect of my operation during the preparatory phases, during which time there was no expectation that we would progress to combat. I was horrified by the driving conditions on the local highways, and concerned by the other risks that were evident. Accordingly, I published my vision: "Everybody goes home". As we entered into the combat phase, I took the time to address every individual in my command, and all those augmentees that flooded in. To them I said, "We will win this campaign. It will be won through the successful execution of thousands of sorties. We will fly hundreds of these. No one mission is so critical as to justify cutting a corner or taking risks." I went on to further emphasize my expectations in this area. In all of my visits and discussions with my personnel, I commented on safety, observed and changed small items to reduce risk, and encouraged them to take action to correct dangerous situations.

A related, but equally critical issue is that you must be mentally and procedurally prepared to deal with significant accident, injuries, or loss. When you get the report of an aircraft loss in combat, it is far to late to be figuring out what measures need to be taken. You need to have carefully thought your way through this eventuality, from the notification of next of kin through follow-up. We have a very structured process for aircraft accident investigation; what will you do in the event of a combat shoot-down? What will your media position be?

To my lasting satisfaction, the entire operation was conducted with no loss of life or serious injury, and no accidents of significance. You must walk the talk.

MISSION FOCUS

While you may believe that impending or ongoing combat operations would serve to focus the attentions of your force, you will be astonished by the myriad of other competing demands. Success or failure will depend on your ability to focus your team on the items that are important, and ultimately on mission success.

In the case of TFA, in the days immediately prior to the first combat sortie, we had so many distractions that it is difficult to remember them all. First, we had just received the advance party of the next rotation (ROTO) of personnel; as it turned out the ROTO would have taken place several days after the start of combat. A team was in from Canada in the midst of a Board of Inquiry (BOI) for the changeover. We were also in the middle of the final detailed planning to temporarily move the entire force to another base for a six-week period of maintenance on the runways at Aviano. In addition, we were attempting to write personal evaluation reports (PER) for all the personnel. It is difficult to imagine a more distracted and overtasked group.

As I listened to the Secretary-General of NATO announcing the commencement of combat for the next day, I realized that I had allowed the situation to go on too long. Accordingly, I issued a one-page order entitled: "Focus on the mission." In it I directed all activities that were not essential for mission success to cease. I specifically halted the ROTO, ordered the BOI to cease and the members to report for duty with my force, ceased the planned move, stopped the work on the PERs, and effectively cleared the decks for combat. I can still recall the incredulous expression of the officer in charge of the BOI team when I ordered him to return to Canada.

Your folks can accomplish miracles, and will put in incredible hours. It is up to you to ensure that every drop of productivity and sweat is directed to the success of your mission. To do otherwise is to invite failure. Focus them on the mission.

RELATIONSHIPS WITH HHQ AND HOME UNITS

Warning: this section addresses the one area in which our training is seriously, and potentially debilitatingly at variance with operations. Our training for Air Force units includes numerous deployments for training operations at away bases. In all cases, the mounting wing retains the responsibility for support. When you need a part, or have to send someone home and need a replacement, you call home base.

When you are operationally deployed, your status is dramatically altered: you are under operational command (OPCOM) to the Deputy Chief of the Defence Staff (DCDS)* under our current construct. That is, you are under the full and direct command of the DCDS, and have no further linkage with your home base. This means that all those normal issues, from spare parts to personnel issues, must be dealt with through the DCDS staff. This is a very good process, as the DCDS has recourse to all assets and every person in the CF. You are on the A-team, at the pointy end, and they will rip their sweater to get you everything that you need.

Your staff, however, will arrive with absolutely no experience or training in this new relationship. Their skills will lead them to contact the home unit with their issues. This is wrong and dysfunctional. The home wing does not have access to the full supply system for parts, the full spectrum of personnel of the CF to fill your need, the full spectrum of airlift to deliver the goods, etc. In fact, they may knee-jerk in response to a call, and a totally inappropriate response may be generated.

In one case, I was considering a request to expand our operations from precision guided bombing to all-weather non-precision bombing. I asked my staff to confirm that we did not, in fact, have any Vertical Ejector Racks (VER) in stock with our task force. (these are needed to permit more

---

* Note the DCDS organization was dismantled in spring 2006. Canadian Expeditionary Force Command (CEFCOM) is now responsible for overseas operations.

bombs to be mounted on each position on the wing of the aircraft; if you are not using precision bombs, then more is better) The first phone call by my immediate staff resulted in a shout of "do we have any VERs" in one of the shelters on the flight line, at which point someone shouted back "do we need VERs? I have the home wing on the line." You guessed it, the Airbus happened to be on the ramp at the home wing, and within minutes 50 VERs were loaded. (several hundred pounds each) Within hours they were delivered to us; success you might say. Not so fast. This poorly staffed and uncoordinated action resulted in uncertified and not-maintained VERs being shipped to us   In the end, they could not be restored to flying status with the personnel and equipment available to our task force, and all had to be shipped back to Canada. A couple of days later, subsequent to a formal request for VERs, staffed through the DCDS staff, we received the needed gear, ready for flight.

I had to continually address this, and briefed every newly arriving individual. I had my chief of staff personally conduct remedial training to further reinforce the lesson. Despite all, we had to conduct specific training for every newly arrived individual to avoid further errors.

PLANNING AND CONFIDENCE

Our peace-time experience leads us to pursue the 99 percent solution. No process is sufficiently efficient; we are consumed by the effort to make things better. This is a luxury that you cannot afford in combat (and one could argue you should not afford in peacetime either).   The search for perfection can lead to mission failure; good-enough has to do. I summarized it as: the 80 percent solution today has a better chance of success than the 95 percent solution that arrives late. To this I would add the corollary that "you are better than 80 percent!" In my experience, no matter how complex the issue, we can rapidly arrive at very workable solutions to the most difficult of problems. Accepting that this will have to do, and confidently proceeding is crucial to success.

YOUR PERSONAL WELL-BEING

We are all raised and trained, as future commanders, on the principle that the leader pushes the hardest, endures the most privation, and sets the example for his force. The image of a selfless commander, dedicated to the men and women of his force, is valuable and important. You must, however, understand that this has its limits in application, and that if you fail as the leader, your force fails.

Physical fitness: I cannot overemphasize the importance of you taking all possible measures to prepare yourself for the possibility of combat. You will be faced with impossible levels of stress, fatigue, and mental pressure. The first step to successfully dealing with this starts right now. You must deliberately take action to raise your physical fitness to a high level. You should adopt the attitude that you are in training for a serious, physically demanding competition. In combat, this initial fitness will erode as you are faced with sleep deprivation, poor eating, worry, coffee, and other stressors. The higher the level you start at, the longer and more successful you will be at resisting the effects, and continuing to function at the required high levels.

In our peacetime existence as officers, we have somehow come to accept the image that the more senior the officer, the less important it is to do fitness training. While it is encouraged for our junior officers to take time during the workday to train, it is almost dimly viewed for senior officers to pull themselves away from the inbox and e-mail. I believe that a compelling argument could be advanced that the more senior the officer, the greater the requirement for dedicated fitness training. The young are naturally given greater resiliency. I challenge you to review your priorities, and to make time in your duty week to get into shape, and stay there.

Sleep: In combat, your folks will be pumped and roaring to go. Their dedication and focus on the mission will lead them to dedicate every waking moment to their part of the mission. In the face of the never-ending barrage of issues to address and meetings to attend, as a commander you will find sleep is a precious commodity. You must deliberately take concrete measures to ensure that you get adequate amounts of sleep to function effectively.

You will find, once again, that your training and beliefs are somewhat at odds with your needs. The image of the leader being the hardest driving, most selfless individual in the force can easily lead you to neglect your own condition. You must fight this. Imagine a commander with a pillow in his staff car; tough to see is it not? This is but one trick that I commend to you. Have a folding cot and sleeping bag installed in your office, and periodically inform your staff that you are going to sleep. They should hold the phone, intercept any disturbances, and allow you to take advantage of that lull in your afternoon. Keep in mind that, whereas they have shifts and time off, the commander is always on!

In my experience, the first few day's sorties were all flown under cover of darkness, to reduce the effectiveness of the air defences that my pilots were facing. After monitoring their safe progress through the enemy zone and back into the safe area from the Air Operations centre, I leapt into my car and drove the two hours to Aviano to be there to welcome them safely back. As I returned in the wee hours of the morning, prior to the next early morning brief, I concluded that I owed it to these valiant young men to be on hand to welcome them back after every mission; how misguided can you get! I owed it to them to be as sharp and capable as possible, to give them the best possible chance for success and survival.

After the second night, in which the sorties were flown even later, I was still running on adrenalin. After the first three days, I realized that I had to get sleep. Accordingly I found the discipline to lay in my bed for eight hours on each of the next two nights; alas, my body was so wired that sleep eluded me, other than a couple of hours each night. By the fifth day I was starting to have physical symptoms, and could no longer sleep normally. I was going to fail. Realizing the danger, I reported to my doctor and he prescribed a couple of sleeping tablets for the next night. One huge night of sleep and I was back on top.

To ensure that I did not get in the same situation again, I was determined to institute control. Having noticed how fatigued (ugly) all the senior officers had become in the early days, and being an engineer, I created a management tool. I printed a chart with days along the bottom, and UGLY as the vertical scale (0 to 8). It was entitled "Commander, how do I look today." Realizing that I could not, in all fairness, ask my staff to comment on how I looked, I devised a formula for calculating the vertical score for each day. Given that eight hours of sleep is a fine amount, I decided to subtract the actual sleep, and plot the result. While this started out as an act of humour, I found it to actually be of great value. I scrupulously added every half-hour in the office, 15 minutes in the car, and hour in bed, and plotted the result. With normal days running from 1 to 2 on the ugly scale (6 to 7 hours of sleep), I quickly recognized when intense operations were causing sleep deficits: the chart started to show ugliness levels of 4 or more for several days in a row. When this happened, I made sleep the priority it had to be, and drove the score back down. Do not let a misguided sense of self-sacrifice allow you to run yourself down and compromise the effectiveness of your team; their lives depend on you being at your best.

Ethics

Much has been made of ethics and the requirement for enhanced ethical training during the past few years. The training, in some cases, has focused on complex situational issues in which individuals are faced with impossibly demanding choices involving negative consequences for any decision. While I support this on the basis that it is intellectually stimulating and causes us to reflect on our core values, it is not truly reflective of the types of situations you will face in your operations. Accordingly, I have condensed my thoughts on this topic into a sort of guide to practical ethics.

How do I define an ethical challenge? I define this as a situation in which you are facing the internal turmoil over a decision to act. While you know what you should do (the right thing), you are under pressure to act against your values. This may be temptation (personal gain, pleasure), orders from your superior, rules, or any other source of influence over your actions. I can generally tell when I am facing this, as I find myself returning to the problem over and over again. In the shower, in bed before trying to sleep, on the way to work, while running; the issue continues to haunt me.

As a lieutenant-colonel, I had the formative experience of being placed under immense pressure to act against my values. In this case, I was pressured by several levels of command above me to alter an estimate, so as to favour the deployment of forces into a combat zone. I had personally led a team through the evaluation of the options, and clearly understood that the aircraft were completely unequipped to survive in the threat situation that was present, and the crews were manifestly unprepared to operate in the region. My strong recommendation had been against the proposed deployment. After considering the initial, subtle feedback from the higher review of my estimate, I began to worry that I might actually be told to alter the estimate. What would I do? I was about to be tested.

Several anxious hours later I found myself in front of the general, and to my horror he indicated that he, the Commander of the Air Force, and the Chief of the Defence Staff (CDS) all wished the estimate to recommend in favour of deployment. Horror of all horrors. What to do? In a flash my decision was made. I told the general that I would completely rewrite the estimate to strongly recommend the deployment of the aircraft, for his signature. At this point he back-pedalled, and said that no, this was to be my estimate. I then indicated to him that, if it was to be my estimate, it stood as written. I was then dismissed.

You can imagine my distress in the ensuing days and weeks. As strange as it may be, six weeks later a similar type of light twin aircraft, operated by another nation, crashed into a peak in the same area, killing a US secretary of state and all others aboard. The resulting inquiries revealed not only the manifest lack of preparation for these crews to operate in this area, but also showed dramatic levels of inappropriate pressure by significant levels of command to deploy this aircraft and push the ill-prepared crew to undertake the mission. I do not sleep with that on my conscience.

My suggestion to you is as follows: when you find yourself continually drawn back to a decision, over and over again, carefully consider whether you are under some pressure to act against your values. Treat this as an alert to you to fundamentally re-examine your situation. Clarify why you are conflicted, examine the source of the pressure, and do what you know to be the right thing. To your own self be true.

A word of warning, do not confuse a difference of opinion with your boss as to the best plan with situations that reflect true ethical issues. You may feel strongly about your proposed solution, but you owe your boss the loyalty to accept his orders and get on with the job in the fashion that he has decided to proceed, to the best you are able. He will frequently know more than you do, and may well have considered and weighed factors differently than you. Your turn to lead will come. This is not to be confused with the pressure to act counter to your internal values.

Leadership and motivation

While every situation will be different, it was my experience that your folks will be immensely motivated in the face of operational duties, particularly so in the face of combat operations. For many it will be their first

CHAPTER 5 — IN HARM'S WAY

experience on operations, and will be their opportunity to see if they measure up. Your task as a leader is thus rendered far easier than might otherwise be the case. Fundamentally, they all know their job; all they need is a clear understanding of where you need them to go as a team, and an underlying feeling that you care about them and value their contribution. Being at the head of such a group brings an unequalled sense of satisfaction and pride.

I can conceive of situations where the operation is sufficiently unpopular, or the sense of mission worth of the group is so low, that the leadership challenges could be insurmountable. Fortunately, I believe this to be an exceedingly rare circumstance.

In my case, the tragic humanitarian situation had been extensively covered in the media, and the reason for us to act was well understood by all. I did conduct several large group gatherings to say a few words of encouragement to my folks, and to answer their questions. That having been said, this was the first combat experience for most of them; they were seriously pumped up already. The vast majority, myself included, had never expected to be called upon to go to war for our country. To be called upon was a vindication of our life choices, and a validation of our training to date. My challenges were to ensure that this supercharged team had very clear directions, to encourage and reassure them as they faced the inevitable initial stage-fright and butterflies, and to step out of their way as they charged forward to accomplish the mission.

Your challenge is to bring every individual on your team to understand and believe that you truly care for him, and desperately need his contribution. Only when every single individual is pulling at 150 percent will your group achieve its best effort. As I have posted on my wall for years, "You can make them work overtime, you can make them work weekends, but you cannot make them perform extraordinary work. They have to want it!" Your challenge as the leader is to nurture this desire to excel in every one of your troops.

Media

While we have made dramatic strides in advancing the level of awareness of media and the role that they play in your operation and the national support for your activities, we have some ways yet to go. Seek out opportunities to further your understanding of the media, and particularly, to expose yourself to media training.

During 1997, I attended the U.S. Air Force Air War College. The year of intense studies included many core topics, coupled with a broad spectrum of electives. From the list of electives, I happened to choose the less-than-popular elective on Media. In the ensuing weeks I was subjected to my first television interview, first radio interview, and finally, my first negative press conference, complete with demanding and pushy reporters out to show me in the worst possible light, and a lady that went somewhat hysterical in the middle of everything. This was anything but fun.

During a fighter pilot's training, we send him to Red Flag or Maple Flag in order to experience his first 10 combat missions during training. He is subjected to the most realistic combat environment possible. The requirement to do so was based on the observation, over the years, that a dramatic percentage of all combat losses occurred during the first 10 combat missions; those that survived had a very reduced loss rate.

For me, the media course was like having the first enemy shots occur during training. Two years later I found myself, standing in a muddy field in the wee hours of the morning, being interviewed by CBC, live on "The National." I could not have been more thankful for having signed up for the earlier training. Make sure you are ready.

It is clear that Canadians will have a significant interest in your mission, and will wish to see the human side of your endeavour. It is equally clear that if you present such a negative image, or if you exclude the media entirely, then the coverage will inevitably be less than flattering. This has two significant and direct impacts on your mission success: first, you have been assigned the mission as a result of a decision made on the part of all Canadians to intervene; should your mission be portrayed in a sufficiently negative light the national support to continue may fade, and you may be withdrawn without succeeding in your mission. Secondly, in the age of advanced communications, your troops will be directly exposed to the coverage back home, both through telephone calls to the family and though internet access to the news coverage. If the coverage is negative, you can bet that this will have a serious impact on the morale and effectiveness of your folks, and may jeopardize your mission as a result.

This should lead you to the conclusion that you will need to have a proactive and well-considered media plan in place, before it gets interesting. You will need to ensure that you have adequate professional public affairs officers on staff, that you have carefully considered the message that you wish to transmit, that you have decided on a policy and process with respect to embedding media with your force, and finally that you have carefully considered the conditions under which you will allow your folks to be interviewed.

This latter issue is fairly important. In general, your folks are actually very good at drawing the limit with respect to the type and detail of information that they give out; this should not be surprising, given their risk should they compromise mission-sensitive information. The issue becomes more one of protecting them and their families. Given that our forces are relatively small in number, and are generated from a relatively small number of bases, it is fairly easy for the media and others to establish the identity of the individuals. This brings the risk of reprisals or threats against the families back home. Again, this can be of sufficient concern to your folks to represent a threat to your mission success. The obverse of this issue is that the anonymous interviews, or the hidden faces approach lead to a sense of a group that are engaged in some dishonourable endeavour. The bottom line is that you need to have sorted this out well before you find yourself in action.

RULES OF ENGAGEMENT (ROE)

In a combat environment, you are faced with the likelihood, perhaps even certainty, that personnel under your command will apply lethal force to accomplish the mission. This is legitimized through the issuance of rules of engagement. These rules describe the circumstances under which you and your unit are authorized to use force, whether lethal or non-lethal.

In my experience, the complexity of the ROE is inversely related to the likelihood that your unit may be called upon to actually use force. That is, in a peacekeeping scenario, where it is highly desirable that force only be used in a last resort, the ROE are complex and highly restrictive. In the circumstances of combat, where the mission expressly calls upon the use of lethal force, the ROE are simple and straightforward. In fact, in the latter situation, the principles of the Law of Armed Conflict become the primary directive for the use of force.

For example, during the two months leading up to the air campaign, we were tasked to fly missions in support of the peacekeepers on the ground in Bosnia. We were to drop precision bombs in support of peacekeepers, in the event that the local situation had severely deteriorated and such dramatic levels of force were necessary to retain control of the situation. The possibility that we might actually have to do this was remote. As a result, the ROE that applied were exceedingly complex, with a multitude of procedures required to authorize the use of force, and complicated measures on the aircraft to ensure that weapons were not inadvertently dropped, etc. In fact, the complexity was of such magnitude that I established the second of my two missions objectives as "No unauthorized bombs."

As it became increasingly possible that we would be engaging the Serb forces in open combat in The Former Republic of Yugoslavia, I sought and received ROE for that mission. These were exceedingly clear and very non-restrictive. In fact, particularly in view of the significant emphasis on collateral damage, the principles of the Law of Armed Conflict (LOAC) became primary in the planning and execution of the missions.

As the commander, you are ultimately accountable for all the actions of your force, most particularly for their application of lethal force. You must take all measures necessary to ensure that your folks are fully aware of all of the applicable ROE, and your expectations of them. In my case, I personally briefed all my pilots prior to their commencing combat flying, and spent considerable time in discussion with them. I was questioned with many "what if's" and was able to clearly communicate to them my expectations. In my case, I clearly and repeatedly stressed the importance of them being certain of their aim point before releasing weapons. We were to bomb targets that were in close proximity to civilian populations, in some cases embedded in cities and towns. My contract with them was that if they were certain, but in error, then I would answer for the results. If they failed to be professional, particularly if they dropped their weapons without having positively identified their aim points, they would answer.

In the event, we had numerous sorties where the pilots flew for hours, faced the enemy air defences, arrived at the target area and could not clearly establish their aim point. They all showed the discipline to hold on to their bombs, swallow their disappointment, and fly the hours required to get home, having failed in their particular mission. I cannot express the pride that I retain to this day in the professional and disciplined fashion in which they performed. I do not believe that the same result would have occurred without my clear direction and communication with them.

YOU AND YOUR LAWYERS

You will be fortunate to have one or more lawyers deployed as part of your team. These talented Assistant Judge Advocate Generals (AJAG) are provided by the JAG, and represent a source of invaluable advice on a wide variety of issues. You must be aware, however, of a number of significant special factors that influence them, and their relationship with you.

First, you will require these lawyers to deal with a huge spectrum of issues. These will range from disciplinary issues with your troops, local contracting issues, interaction with local national courts, real estate leases, LAOC, ROE, your orders, and a plethora of other issues. How can anyone be so competent you ask? The answer is that they cannot be, thus the first justification for the unique command relationship that exists for your AJAGs alone. They are the only individuals that will be present in your command that are not under your direct command. They are responsive to you, but remain under the direct command of JAG. This allows them the right to directly communicate with the JAG and the large and diverse legal organization back home. This affords you the benefit of the very best of possible legal advice.

The next point is somewhat more subtle. As the commander on a deployment, you will be given the most authority that can ever be bestowed upon an individual. You are literally given the power to decide life and death, and all lesser things. This is an exhilarating and frightening proposition, one that brings with it the greatest of satisfactions as the mission progresses. As we have learned through past events, this absolute power has led to abuses when the talents of the commander were less than that required for the situation. While all subordinate officers have the ethical responsibility to act in the face of wrongdoing, this is a particularly demanding challenge in a combat environment. Most of us have read of several famous cases, such as the "Mutiny on the Bounty" where subordinate officers have faced the supreme challenge of taking action in the face of commanders that were failing. This is a demanding act that is rendered doubly difficult in view of the chain of command relationship, and the necessary respect for authority that is ingrained in us all from the first day of basic training. By placing the AJAG outside the chain of command, at least one individual is placed in a position from which it would be somewhat easier to act in the face of wrongdoing. The AJAG acts as a safety measure, to guard against a commander overstepping his authorities. This is a right and necessary thing.

You will need to be aware that AJAGs are not trained in the art of military command to any significant extent. While they are consummate experts in a number of areas of law, and in their relationship with JAG and his staff, their military training may range from a reasonable amount to a relative light dusting. They will provide you with high quality legal advice; their responsibilities do not include the provision of military advice, whether concerning the conduct of your campaign or leadership in general. Be cognizant of this, and be prepared to assist your AJAG in understanding the scope and limits on their responsibilities.

Finally, and most importantly, you must be absolutely clear as to your relative roles, responsibilities, and accountabilities. Your AJAG is responsible to provide you with legal advice. He or she is absolutely accountable for the quality of the advice provided, and can have their professional accreditation revoked for poor advice. As the commander, <u>you are absolutely accountable for your decisions</u>. In a proper relationship, you will be provided advice, couched in terms such as "it is not advised that", or "I recommend that". Should you hear "you can't", "you must" then you may be straying from the appropriate relationship. You must not look to your AJAGs for decisions.

COMMAND ACCOUNTABILITY

Finally, as the commander, you must remain cognizant that you are ultimately responsible for everything that your force does – absolutely everything. While I was faced with circumstances where the coalition leadership assured me that they would personally take responsibility for the results of proposed attacks or tactics, it was absolutely clear to me that I would personally answer for my force and their actions. NATO or some other coalition does not go to court – you do!

This led me to refuse some missions and to reject some tactics. In retrospect, the events vindicated my stand in each case. I will cover two, in brief, to illustrate.

The first was a tasking to attack a significant communication centre, one afternoon. The centre was a legitimate target with great importance, and was assessed to have a low probability of collateral damage. The LOAC principles would allow attacks up to and including high probability of collateral damage for such a target. Upon closer examination, my planning team indicated concern to me. They had observed this target on a previous mission, and knew that it was surrounded by immense parking lots. The size of the facility indicated the presence of hundreds of civilian workers. While civilian workers involved in providing military capabilities are legitimate targets, the attack date was during the days immediately following the inadvertent attack on the Chinese Embassy. I approached the NATO commander and was initially rebuffed. After being directed to the targeting cell, I was again told that the target had a low collateral risk. I asked why this target was tasked in the mid-afternoon, assuring a large work force and many casualties. There was no answer. Accordingly, I refused the target. The next day the chief of the targeting cell approached me; the target folder now had a unique red page that stated that this target was only to be struck in the early morning hours. A couple of days later we were again tasked, and eliminated the target without causing the elevated level of collateral damage and subsequent adverse public attention that would surely have resulted if we had performed the mission as originally tasked. In view of the furor that arose over the Chinese Embassy, had I not objected, we might have created such debate that the entire campaign could have been placed in jeopardy.

In a second case, the Air Component Commander (ACC) was placed under tremendous pressure to increase the weight of attacks against the Serb Army in the Kosovo province. In view of the shortage of airborne forward air controllers, to find and identify legitimate army targets for the attacking fighters, he decided that he would allow flight leads to over-fly the area and find their own targets. I was concerned, as the fighter aircraft were not equipped with the required capabilities to effectively find targets, and in particular, to positively confirm that they were legal military targets. Despite the personal assurances that the ACC would "personally take full responsibility", I issued direct orders to my force that prohibited this procedure. Two days later a pilot of another force incorrectly identified a tractor with a trailer full of refugees as a target, and attacked with deadly results. I felt sympathy for the pilot involved, particularly as he was personally paraded before the media to explain his actions; he will live with this decision for the rest of his days. The next day

I removed my prohibition on such tactics, as NATO issued its own restriction. Fundamentally, as the commander, you must be absolutely cognizant of the fact that you, and only you, are ultimately responsible.

HONOURS AND AWARDS

This is one area of primordial significance, yet one that is the most neglected. In an era in which all talk seems to have shifted to pay and allowances, and "Queen and Country" is disparaged, you will be amazed by the extent of the emotion that is associated with recognition. This is an area in which the wheels of our current system do grind slowly indeed. You hold a sacred responsibility to your folks; they will face great danger and hardship under your command, and you are accountable to ensure that their sacrifice is recognized. This is far from simple; our system is fairly complex, the application process is long and less than rapid. Fundamentally, however, the entire outcome is based on your early and continual engagement on this critical topic.

In my case, it took five years of ceaseless efforts, and the support of a number of key senior leaders, to finally achieve appropriate recognition for my warriors. This did pay off for future missions, however, so possibly this situation will be minimized in the future.

The bottom line is that you need to be fully aware of the spectrum of individual and group recognition and honours, and diligently apply your energies to ensuring that the deserving folks are recognized appropriately. This responsibility does not end with the end of your mission.

CONCLUSION

In reviewing the notes above, I have the sense that I have painted a portrait of command in combat as a severely demanding role that will push you to your very limits, and will present you with the greatest challenges possible. If so, I have succeeded in my aim. You should not shy away from such an opportunity; remember that the greatest satisfactions are found in achieving success in the face of the greatest challenges.

An operational command will represent the pinnacle of your career, and prove to be the validation of a lifetime of dedicated training and development. Good luck.

CHAPTER 6

OPERATION APOLLO

Lieutenant-Colonel G.L. Smith

Let's step back to what has now become a notorious day in modern history: 11 September 2001. Each of us can vividly remember what we were doing that day when those historic but tragic events took place in New York City. For me, I was scheduled to conduct duties as the Aircraft Commander for a routine local flight using the C-130 Hercules transport aircraft, engaging in a 'hostile theatre operations' training exercise in the local Trenton flying area. As Commanding Officer of 429 Squadron, I recognized that this mission, like all of the others that came before in my twenty-three years of previous military flying experience, was designed to enhance the readiness of assigned crews to rapidly respond to Canadian defence priorities. Little did I recognize then how relevant those years of international training, humanitarian and peacekeeping operations would become.

My aim with this chapter is to share what I believe to be the relevant lessons learned through my command experience conducting air operations within a hostile theatre. In particular, my task as the first rotation commander of the Tactical Airlift Detachment for Operation (OP) APOLLO will be examined to fulfill this aim. By conducting this detailed examination, I hope to make a direct contribution to those who may find themselves confronted with comparable circumstance in the future. Initially, I will present an overview of the operation's context and history through the medium of a typical day, followed by a focused retrospective of my experience with a detailed emphasis on leadership lessons learned. Finally, I will conclude with a summary of the most substantial lessons I learned that can directly prepare others for similar leadership challenges. By using this approach, the intent is to share the layers of experience I took away from this deployment, but in doing so, perhaps I might subtly entertain along the way.

THE TASK REVEALED - A MEMOIR

Again, I remember 11 September 2001 with lurid detail. Like many, I met the televised images of the catastrophe with disbelief and cautious skepticism, viewing them at first as one might watch a sensational 'Hollywood' movie trailer. During the immediate aftermath, I knew little of the planning that was taking place nationally; however, I intuitively knew there had to be a response. Fortress 'North America' had been breached and Canadian sovereignty had been directly threatened. In the weeks that followed, I can clearly recall the Prime Minister, the Minister of National Defence (MND) and the Chief of the Defence Staff (CDS) jointly announcing Canada's commitment to the newly declared 'War on Terrorism'. For the Department of National Defence (DND), this proclamation amounted to the planning for and execution of OP APOLLO – Canada's contribution to a coalition-based, multi-national military mission prosecuted in a theatre of operations then considered to be a targeted centre of gravity for international terrorism: Afghanistan.

I received a telephone call during the afternoon of 8 October 2001 from the Wing Commander of 8 Wing Trenton. He informed me that I had been chosen to mount a force of approximately two hundred personnel and three C-130 Hercules aircraft for a six-month deployment to an undefined location. Suffice it to say that my small part in the world and my family circumstance were turned upside down that day. In the days and weeks that followed, this scene would repeat itself across the country in the households of many service personnel called to duty for OP APOLLO. For three months, we prepared, and the team trained individually and collectively. Throughout the period, the main deployment body and a proportionate back-up team held a twenty-four hour standby posture that was emotionally exhausting for us and for our families. Straddling the 2001 Christmas/New Year's traditional holiday season, the anxiety of knowing one will be leaving but not knowing 'when' was noticeably wearing upon all of us. For those poorly informed service personnel, rumours of departure dates and basing location options were rampant and the 'ebb and flow' of morale changed accordingly.

Contrasting directly with airlift operations experiences of the past where supporting actions had been of relatively short duration, this mission was planned at the outset as a six-month operation – a fundamental paradigm shift for many of the flying crews and support personnel sourced from within the Air Mobility community. This was met with some resistance due, in part, to the 'inertia' of the group's cultural sense of 'who they were' and past practice. Magnifying these already destabilizing organizational forces, an observable amount of paranoia was apparent in smaller but nonetheless appreciable amounts: this mission was not 'routine' peace-keeping or a 'classic' humanitarian operation in any sense. Nonetheless, most of the team was highly motivated by the spectre of terrorism and the practiced techniques of terrorists to date.

Finally, commencing mid-January 2002, following a tense but thankfully uninterrupted holiday season, the Tactical Airlift Detachment was finally given its orders to move. This was a huge logistics event unto itself, but relatively small in comparison to the parallel events that were unfolding across Canada and in many of the participating nations across the globe.

After almost a full day of travel, we arrived at our destination and were confronted immediately with an alien culture and a despicable austerity in what would become 'home' for the next six months. In a few short days, a rudimentary airfield operation was set-up in this remote desert location and our first airlift events commenced. The runway was provided by our host nation, however, the infrastructure from where we would conduct operations was built entirely following our arrival. Due to this deficiency, my first three weeks of functional command in this environment were carried out exclusively on a field message pad. Two days after arriving, the Detachment's aircraft maintenance, traffic and air movements, operations planning, aircrew and support staffs were now executing their respective crafts in order to launch viable sorties. For all of us, the bustle of activity helped to get lingering images of family separation and the tears shed on departure out of our minds (if only temporarily). For most, a typical day was eighteen hours from start to finish. However, as we became more versed in the needs of the mission, the days became more tolerable as we learned our own efficiencies and adapted the organization to meet the changing needs of our complex mission. After all, we were only doing what we had practiced so many times before, or was it the same?

Planning commenced at 2300 hours before the following mission day. There were no days off to the operation per se, but as a general rule, most personnel were scheduled to rest one day per workweek. Assigned flying crews would arrive at 0300 hours daily to commence their detailed study of the planned routing and prepare contingency escape and evasion plans in the event of being forced down over hostile territory (an exercise which I have often referred to as 'angry camping'). All aspects of this planning were based entirely upon the in-depth analysis conducted by on-scene intelligence experts who were highly tuned to the rapidly changing mission parameters and situational developments across the entire theatre of operations.

On occasion, I recalled remarks from retired Major-General Lewis Mackenzie being interviewed on Canadian television immediately prior to our departure during which he mentioned (paraphrased) that 'the Afghanistan national sport is removing expatriate (foreign) troops from its soil' (to note, this was a reflection upon the Soviet Union's unsuccessful attempt to invade in the early 1980s and failed British efforts to colonize the country earlier in the century). It was obvious that there were pockets of versatile, war-hardened Taliban soldiers who did not want any of us in their country, especially when we were seeking to fundamentally depose their tribal way of life, unseat the ruling warlords, and shift the centre of power away from those local and regional benefits that democratic order would deny. The precarious interaction between friend and foe in this upset nation was always observed by me in this context and carefully integrated into my overview of each day's mission assignments.

As the planned departure time moved closer, receiving their weapons and ammunition only heightened the sense of focus for the aircrew. Meanwhile, aircraft maintenance personnel readied the C-130 Hercules aircraft which included: installation of nearly 3000 pounds of Kevlar armor plating designed to protect the crew-members from small arms fire; functional testing of protective decoy flare and radar-confusing chaff dispensers intended to fool enemy missile and radar systems should they exist; and, completion of a detailed pre-flight aircraft inspection. By any other description, it was the ultimate tune-up. Air Movements personnel simultaneously loaded the aircraft to meet critical mission timings. Missing these crucial timings usually meant mission cancellation. Defensive airspace corridor management practices demanded highly accurate arrival and departure times and were critical for defence of our aircraft as they arrived into and departed from all airfield locations in Afghanistan. Although a known air-to-air threat was non-existent, routings and timings were manipulated to assure a degree of randomness to our mission behaviours. The surface-to-air threat and the terrorist threat while on the ground 'making the deliveries' was very real.

We helped sustain the logistics life-line to all Coalition forces in Afghanistan and elsewhere within the theatre of operations by delivering their precious freight daily. Without our sustained effort, in-country units might go hungry, be without the proper equipment to carry out their mission, or not be capable of defending themselves. In the coalition context, the Canadian Tactical Airlift Detachment supported many nations including Canadians, and was very much part of a greater coordinated effort – a perspective that was not appreciated by all within the theatre of operations.

A pre-mission weather briefing is completed, a last minute check of the intelligence picture is updated, and a final crew briefing is conducted. Only then does the crew make their way towards the waiting aircraft. At this juncture, the mission either succeeds or fails: the aircraft has been mechanically prepared and configured for its mission; the freight load is onboard and rigged for air transport; the aircraft's self-defence systems have been checked and rechecked; the checklists commence; and, the engines are started. The aircraft conducts its ground taxi and departs off of the adjacent runway on what will be a fourteen-hour day for them, half of which will be spent over unfriendly territory.

As I listened for the sound of them leaving overhead, I quietly wished them well but there was little time for reflection. As they left, a second mission was being prepared to depart, perhaps to the same destination and perhaps not. I, like others, was already preparing for the next day's missions and carrying out the necessary communications with associated staffs and senior commanders in Canada and elsewhere advising them of our conduct to date. Coalition Headquarters had already completed planning for the next day's sorties. The incessant cycle kept us distracted from the tensions of the environment and the loneliness of family separation.

In time, I was informed that the aircraft and crew made it to their destination without incident. The living conditions at the location of landing are not too far removed from the Stone Age, but they are safe and the precious cargo is off-loaded. After forty-five minutes on the ground, the aircraft and crew departed to come home. Again, they transited hostile territory ever vigilant of their surroundings and wary of the

CHAPTER 6   IN HARM'S WAY

incessant chatter over on-board radio systems and the constant input of defensive sensors picking up the signals of friends (or foes). Finally, they went 'feet wet' – a phrase used to describe being off of the hostile land mass and safely outside any enemy engagement zones. The mind says it's time to rest for the crew flying the mission. For them, it is now hour twelve of a sixteen-hour day and the stress of being exposed to mortal danger causes the mind and body to feel that it has been exposed to a day two or three times as long. Temperatures often exceeding 50 degrees Celsius, frequent desert sand and dust-storms amplify the effects of fatigue for everyone, to include those like myself who wait for the aircraft's return to base.

When the crew returns safely to home base, we are all glad to see them safely back. The aircraft is back and no damage from hostile action is found. As one young fellow on the maintenance team pointed out to me on one occasion, it's a good thing they are back because he was looking forward to a vengeful game of ping-pong with one of the crew-members that same evening following a sound thrashing at our small recreation facility the night before. In just four short hours, the cycle would repeat itself as it would for the next one-hundred and seventy days. I find some small comfort in knowing that another success is in our hands.

The entire operations team would discuss the day's events in detail following an exhaustive crew debriefing. We would then turn to apply the lessons learned from our mistakes and commence preparations for the next day. After a terrific meal prepared by Canadian cooks on-site, a quiet evening of watching a movie or exercising, it was off to bed to challenge another day. Perhaps it would be another destination and certainly another freight load but there was always the uncertainty of not knowing where the 'bad guys' were, always hovering in the background of my thoughts. Constant exposure to the burden of this uncertainty was a trial of the team's spirit and morale just as it had effect on me. For our Detachment's rotation, this sequence of events took place 176 times. In total, the Tactical Airlift Detachment moved 4,400 passengers and 4.5 million pounds of freight in 327 incident-free missions. Over 90 percent of the missions we were tasked to conduct were completed as we say in the airlift business 'on target and on time'. Without a doubt, these achievements exceeded many of our performance expectations going into this harsh and uncertain environment but it had even greater effect on all who took part.

LEARNING LEADERSHIP LESSONS

In the preceding section, a brief description of the operational tempo and routine was given; the environmental circumstance and the situational variables that challenged the Tactical Airlift Detachment were illustrated; and throughout, some personal reflection was injected for detail. In this section through the examination of narrower situational perspective, more subjective detail regarding personal leadership technique will be presented. As one reads this next section, it is critical to understand that I believe that leadership, for all of its facets and supporting behaviours, is not just a simple application of lessons learned by rote technique. One's own leadership style will evolve over time as other role models are emulated through the observation of charismatic and potent (or even deficient) leaders. Personal leadership education is a highly subjective and very personal developmental event. I believe learning leadership skills bears highly upon and directly interconnects with one's personality and unique work experiences, entrenched value-sets and practiced ethical conduct, and, is best tested under the rigors of temptation, influence, and changing circumstance rather than the classroom. In short, it is hard for someone else to stand completely in 'my shoes' as each situation is described in this essay; however, notwithstanding this limitation, I hope that sufficient understanding will remain to recognize and understand why I acted as I did.

Preparing the Battlespace

Instead of considering this section as a practical application like the title might suggest, consider 'preparing the battlespace' in terms of developing effectiveness through individual or group competency and collective trust-building within the team. As a follower in the varying leadership styles to which I was exposed in my past, I developed the sustained belief in the need to establish solid competency in my military occupation. Exercised through my own uncompromising drive towards being the very best at those duties I was trained to perform (i.e. a C-130 Hercules pilot), I actively learned, listened, and put in place over many years the improvements I needed to secure habitual skill-sets that I could rely upon in the most trying circumstances. While some of the 'best' leaders to whom I have been exposed have not always had the 'reputation' of being the best in their respective occupational specialties, I understood that for an 'ad hoc' group like the Tactical Airlift Detachment gathered together for a highly challenging purpose, being 'known' as a good 'operator' would help me by providing some start-up inertia to at least initially influence my chosen subordinates. For those who might challenge this proposition, consider the daunting challenge of building a trusting team under similar conditions when competency may be suspect. My first lesson here is to work continuously towards self-improvement in your chosen occupational field. Sophistication in other areas outside this limited focus will come with time and experience, but the first step towards becoming a respected leader, in my opinion, must be the establishment of your own core competencies. This learning must never be permitted to stop.

Unlike many Land and Maritime combat units where they are structured and manned for self-sufficient operation to varying degrees, the formation of the Tactical Airlift Detachment was an 'ad hoc' event. In a normal garrison state, Air Transport Squadrons generally do not have the organic capability to carry their own maintenance, air movements (the Traffic function), and traditional 'base-level' support functionality into the field (e.g. administration, communications, technical support services, finance, food services, transport, construction engineering, etc.). These functions, at the home base, are provided at the base or 'Wing' level. In the case of some Air Wings, like 8 Wing Trenton where I was stationed during OP APOLLO preparations, the aircraft maintenance function is also centralized instead of being nested within the operations unit – an efficiency-based organizational model designed to rationalize four collocated C-130 Hercules Squadrons at that location.

For many years prior to OP APOLLO, the Air Mobility community and the former 'Air Transport Group' had been accustomed to operating from 'Airlift Control Elements (ALCEs)' which were not entirely self-sufficient entities. I had personally functioned as both 'operations planner' and 'crew-member' executing in a number of past events that included numerous deployments supporting the NATO Northern flank 'Express' series of exercises, humanitarian relief in Somalia, airlift support into the former Yugoslavia and other locations across the world. For me and for the team I had to build, self-sufficiency planning was something for which I was neither well prepared nor experienced to fully consider, hence I was compelled to rely upon surrounding 'experts' to permit me to function beyond my personal boundaries of comfort and competency. As a leader in this new circumstance, I found myself forced to trust the advice of my chosen team but I want to emphasize that this was not an unpleasant circumstance: it was the advent of the team's emerging group trust. To enable this leadership transition, I was given the latitude to choose my immediate supervisory team. Ultimately blessed with 'the pick of the litter' by selecting those with praised reputations and proven competencies, I relied heavily upon their opinions in their respective occupational focus areas in order to begin the building of a greater team; however, bringing these personnel together brings new challenge.

The advice I can offer in this context is that while one may witness the dynamic of differing personalities operating in close proximity to one another in a strained environment, one must: carefully adjudicate over any differences that may exist; intervene only when absolutely certain (debate and argument – sometimes called 'storming' can be a healthy group formation exercise at the outset but is not healthy for any group over a prolonged period); and, remain cautious in alienating opinion by dithering in one's own decision-making or constantly defeating input from one or several of your immediate subordinates. Whether in the planning or execution phases of a deployed operation, gather around you the right people with proven skill-sets to help accomplish the mission. By doing so at the supervisory level, the competency of the entire leadership of the Detachment will never be in doubt – an essential step to building a robust and effective deployed force. Only then will the followers follow. Using the analogy of building a fire, one must first accumulate the 'kindling' (i.e. the right people), and then continue to provide the fuel to assure the fire does not extinguish. In this regard and relevant to my next point, that 'fuel' is trust.

I would suggest that trust-building is an elusive human event to define in quantitative terms and even more problematic to achieve with any certainty in a short period of time. Without an exhaustive or more formal analysis of its complexities, I have found trust involves the complexities of human emotions and behaviours on an individual level, competing priorities of both individual and group goals, and the interactions of internal and external influences on those behaviours and goals. As a leader, the challenge is to quickly find a mutually satisfying balance between these competing factors and to build trust throughout the team, which is really in my opinion, the 'output' of that balance.

In a garrison context, it is simple to illustrate that internal and external influences are limited and relatively predictable. Personnel are in familiar work circumstances and the manner in which they are accustomed to performing their duties is routine. Removing them from routine by displacing the location and the circumstance will be destabilizing enough, but removing them in such a way that their home circumstance is impacted brings another dynamic to the argument. In my third lesson learned of this chapter, I will provide an example of how I built trust among the more than two hundred personnel by addressing what I will casually call 'the enemy at home'.

Before there is any misinterpretation with my closing remark in the previous paragraph, be assured that I believe that our families and their security are the most motivating forces in our lives. Indeed, I consider family welfare and stability as a centre of gravity for success in any operation, and that directly threatening that stable foundation can lead to failure of any organization. Stable families produce and sustain personnel who are not distracted and who may be more energetic as a result of active family support. Our spouses, children, siblings, parents and other family members are not 'enemies' but instead give us the solid foundation from which we all live and grow. This was a perspective that recently retired CDS, General Raymond Henault, maintained as he moved the focus of our obligations as military personnel back into the home and the needs of our dependents rather than being exclusively task-centric with family interests wholly excluded from the development of military policy. As military personnel, we recognize that we have a unique obligation to service that encroaches directly upon the tradition of a stable family foundation and it is to that negative effect of deployment that I offer my next leadership example.

During the process of selecting personnel for OP APOLLO, I observed with renewed acuity that those being selected were being immediately disturbed from the home and work routines to which they were accustomed. It became equally apparent that the comforts of living and working in Trenton, like other communities from where many other detachment personnel came, provided a sedate and relatively undisturbed lifestyle for families excepting the annual 'threat' of a posting to another location. In consideration of some very fresh memories lingering in everyone's minds following 11 September 2001, I became very aware that I had a unique and focused problem of family anxiety to defuse – it was happening to me in my own home. For most families, the thought of a loved one traveling to a foreign location for the purpose of challenging those who would carry out such a despicable act as the attack on the World Trade Center buildings in New York had enormous significance. In many cases, spouses and immediate families were confronted for the first time with the occupational 'fine print' detailing the somewhat unique military concept of 'unlimited liability'. In some isolated cases, social worker intervention was the result for those who could not come to terms with this tension.

As the leader for this deployment, I took it upon myself to resolve these festering anxieties and quash any unfounded rumour – this was, in fact, the 'enemy' against which I had to defend myself and the integrity of the mission. I did not perceive this responsibility to lie exclusively within the domain of each service person assigned to the mission. I needed to demonstrate ownership of the operation and instill confidence in each family member to show each of them that I had the competency and vision to see my way through this confrontational mission challenge.

Like choosing the members of my supervisory team, I sought out the assistance of experts to answer this call. I engaged national-level Personnel Support agents (Canadian Force Personnel Support Agency or CFPSA) who were accustomed to family support for ongoing deployments in the former Yugoslavia. I canvassed for and received the support of area social workers, legal officers, medical experts, the Chaplaincy, insurance and financial counselors, and took advantage of the on-site Military Family Resource Centre to defuse family concerns. Again, the situational circumstance must be re-emphasized: personnel in Trenton were not accustomed to six-month deployments and these mission durations were typically perceived to be within the exclusive domain of the Army experience. Families were a little shell-shocked by OP APOLLO, but putting those 'people' resources (i.e. the experts listed earlier) in place did not automatically draw out the concerns and the questions from anxious families. My solution to this was to conduct, on a regular and recurring basis, a series of open 'town-hall' sessions two evenings per week where all affected service personnel were invited to bring any members of their families along to find out the answers to any pragmatic questions from any of the subject experts but more importantly, to have the chance to speak with me directly. Brief but formal presentations were a matter of course in the information-sharing sessions, but my opportunity to speak to individuals who expressed real, reasonable, and often emotional concerns under the circumstances was the investment in my personal energy that provided the greatest long-term return. I used this chance to educate, remove doubt as to my integrity and my intentions, dispel a litany of rumours and present my most solemn promise to make decisions that would assure the safety and security for those people in my charge whom they cherished. In effect, I demonstrated my focused intent and my competency to concerned families. They, in turn, responded with their trust in me because I showed active interest in engaging their concerns.

Simultaneously, I held information briefings with all of the assigned personnel to the mission during the working week and although the information shared was packaged differently, the aim to building trust was identical. By using this two-pronged approach to address the needs of the serving member and his/her family, their emerging, individual and collective trust in me diluted their level of anxiety and lessened the impact of upset family routine. This technique allowed me to show that I cared and that these people meant something to me, but most instructive in this example is that I did care – absolutely and unconditionally. Upon reflection, the proof here was that no one told me how, when or why to do this, I simply did it on my own because I felt it had to happen.

Managing Expectations

As would be expected, successfully celebrating Christmas and New Year's in the winter of 2001 with family was an enormous joy for everyone. Strategic site reconnaissance was ongoing as select personnel from the National Defence Headquarters (NDHQ) sought out a viable location for the detachment in the shadow of heavy British and American presence in the theatre that had already gobbled up much of the operational theatre's infrastructure and airfields. While the rumours of 'when the team was leaving' diminished with the passage of the holiday season, the informal gossip channels were soon energized with the next most important question on people's minds: where were we going?

It was very difficult to dispel these rumours as I was not aware until a few short weeks prior to the move of our Detachment's reconnaissance party. We knew early on that we were not going to be located in Afghanistan. This provided some relief for all including concerned families; however, emotions were stirred by televised and print media that was rife with rhetoric describing Canada going to 'war'. Reconnaissance teams were in motion to place a Princess Patricia's Canadian Light Infantry (PPCLI) battle-group on the ground in-country, Canadian sniper teams were already actively embedded with United States forces, and American television networks were abuzz with images and stories of the ongoing campaign to flush out Taliban fighters from the cave complexes in and around the larger cities of Afghanistan.

I can recall the atmosphere in and around Trenton being highly charged in the early part of January 2002 as a result of these news elements, and I felt intensifying pressure as our time to move drew closer (upon reflection, I suggest this may have been the burden of command coming to call). Last minute scrutiny from higher headquarters in the form of critical organizational reviews and readiness assessments, ongoing Public Affairs work to support the good work everyone was doing, and local preparations for formal departure ceremonies were dominating my daily agenda in these last few days, but I knew we were ready to move when the order was given.

I refer to this part of the chapter as 'Managing Expectations' because with the acceleration that seemed to take place leading up to deployment of the first personnel from the Detachment, it would have been easy to get caught up in the distractions that would cause one to forget what had been previously pursued in earnest: sustaining the lines of communication between yourself, the entire team and their families; growing the trust that had been brokered weeks and months earlier with colleagues and their families; and, fostering in yourself and your subordinates the expectation and the confidence to succeed. Never forget that the assigned personnel and their families expect you to be there for them right through to the date of departure.

The next challenge, with the assistance of the Detachment's Chief Warrant Officer, was to manage the focal shift from family to task, recognizing that family never drops out of sight per se. This shift does not take place in the departure lounge at the airport, but instead overlays the few days prior to departure and continues into the first few weeks following arrival at the operations base.

Amidst the departure ceremonies, the intensive media presence, and the inevitable shedding of tears of family and others upon separation, the only consolation is knowing that you did your best to get everyone to that point. Rather than sensing support and exhilaration in the moment, the emotion was more akin to loneliness: one intuitively understands at that time what the chances for success will be, but how those successes will reveal themselves and at what rate will weigh heavily upon one's mind. Nevertheless, having done one's job properly as a commander, the trajectory that has been established will be in motion, and the personal energy needed to sustain the group as it moves half-way around the world, away from the people and things that are held most dearly, will show itself just as the first steps are made onto the departing aircraft. One has in place all of the proper tools to do the job correctly, the next step in the process is to make it work at the other end of that long voyage away from home.

Hit the Ground Running

At the close of the preceding paragraph, the word 'trajectory' was mentioned. The departure from home base has taken place, the memories of loved ones are still very fresh in everyone's minds but the transition between the planning and execution phase has now taken place. The rumours have disappeared, anxieties linger, and although the mantle of leadership grows weightier, it should feel as if the team believes in you personally. You have captured their trust and they absolutely believe in the success of the mission. Equally important, you should not see unfamiliar faces as you look at each of them during the transit – a good leader has gotten to know his people. The trajectory has indeed been set. Now all you have to do is let them go: let them 'hit the ground running'.

Having done your job correctly before deploying, you no doubt have spent countless hours building trust with your team. The immediate return for this investment in emotional energy is their trust in you but it is by no means the end of that process. It is now your turn to trust them and this is by no means a small feat.

The burden of responsibility for the lives of subordinates and the intensity of focus on mission accomplishment has enormous potential to change personal leadership behaviour in a field situation. For as much anticipative planning one might have undertaken to prepare oneself for the 'feeling' this burden brings while planning at a home base, the act of moving into an operation location has the effect of hyper-focusing your sense of wanting to control everything and everyone. Whether manifesting itself at the deployed location through countless daily 'Orders Groups', lengthy coordination meetings, or putting in place inordinately bureaucratic reporting practices, it is critical to resist the tendency to over-control those who account to you for their performance and who ultimately will make the mission succeed (or fail). While in the planning phase for the deployment, one may have thought that the trust-building described earlier in this chapter was a one-way proposition. It actually (and quite subtly) was a two-way process. As a leader who fosters the trust of the subordinate team, one simultaneously grows to acknowledge the technical competencies and potential for innovation that all of your team can offer. The net result is that your trust in them grows proportionately.

One must be extraordinarily careful not to allow the intensity of the newly deployed and quite foreign circumstances to lead you to be their 'Mr. Hyde' from the 'Dr. Jekyll' you may have portrayed yourself as before deploying. If you succumb to this very real leadership pressure and fundamentally change your behaviour as a result, trust in your leadership can vanish and the team will struggle. If this happens, the secondary effects can include the likely result that your leadership will have been eroded to the point where a new, informal leader will arise from within the group to indirectly challenge your authority.

What was just described did not happen on this deployment, but I have personally experienced this as a subordinate on other less intense operations. As mentioned in the introduction, a properly mentored leadership style and the lessons learned through experience and professional education put into practice will serve one well in the most stressed circumstances. Do not forget these very personal lessons as you suddenly find yourself under the spotlight of a comparable situation of intense responsibility.

There is also a pragmatic side to your new role as a leader. Not only must you continue to show trust as has been just illustrated, you must provide the means with which personnel can fulfill their mission obligations. Expect struggles for resources and stressed logistics chains to rule the day, particularly when the place one normally considers home is half a world away. One must quickly discern between what is essential and what is desirable through open communications with a well established, well exercised, and disciplined chain of command; however, you must avoid the practical tendency to become 'involved' or 'wade in' to every problem or 'dissatisfier' that shows itself.

The role of the leader to sustain morale is clearly entrenched in every historical account of military enterprise but it is not the personal mandate of the leader to solve every problem nor can one expect any operation to take place without mistakes being made, efficiencies being lost, or problems never being resolved. Solving the big problems is your job. Solving all of the little problems is the challenge for your subordinates in the chain of command to confront. Give them room to manoeuvre and provide them the resources to achieve solutions but most importantly: set attainable goals for them; set reasonable expectations for yourself; and, be prepared for the likelihood of mistakes being made by others (do not forget yourself in this respect either). Steering your way through this particular obstacle course can prove to be emotionally stressful and can lead to psychological injury if not consciously anticipated. It is not a comfortable environment for a perfectionist nor is it for one who is especially quick to show the full spectrum of human emotions. As the leader, you will quickly realize that all eyes will be upon you always. Confront adversity with resourcefulness, oppose obstacles with innovation, and give public credit to those who step up to challenge.

It would have been easy to sequester oneself in an office space. With the 'modern' deployment heavily supported and reinforced with workstations, miles of cabling, and other automated utilities from which to gain relevant information or communicate with others near and far, heavy demands will be placed upon one's time to respond or draw from these persistent systems. Continuously remind yourself that these are tools and not replacements for leadership and the presence you must command and execute among the entire team. Done properly before deploying, your team is well known to you so do not disappear from sight as a result of becoming a slave to the technologies that will inevitably surround you. Form early habits to 'get out' and do not rely exclusively on formalized meetings to achieve that standard – talk to people and talk to all of the people (avoid inadvertent segregation of 'favourites' from the others).

Fitting It All Together

Most of the commentary to this point has been intentionally fixed on the internal interactions between the leader and the team, but as might be the case for future deployed circumstance, there is another dynamic to consider: interoperability and the relationship with collocated units. I will now turn the emphasis to the need to assure open and respected lines of communication within a deployed unit and describe, through example, the forces a leader must overcome to assure that these same lines of communication do not become pipelines for destructive forces that can quickly disintegrate a cohesive team or generate disruptive social energy between units.

Stepping back to the point of origin for the Tactical Airlift Detachment, it was originally envisioned by myself and others in the senior leadership chain to be a self-sufficient, stand-alone unit capable of providing a potent airlift capability in a hostile theatre of operations. As was briefly described earlier, organic elements were put in place for all manner of support that any deployed unit would require to survive. As the commander of this package, much effort was made to assure all participants that they had a share in the operation. I worked very hard to reinforce the notion that administration clerks had as much role to play in the mission's success as the pilots flying the missions but the personnel themselves had to make this concept work. By effect, the detachment's successes (and failures) were owned by none but shared by all. The following is one example of what needed to be overcome to achieve this desired outcome of 'operational ownership'.

The familiar garrison circumstance of 8 Wing with the Maintenance teams organizationally separated from the flying crews was now foreign to a team placed together in the same physical proximity under the same closely-knit organizational structure. Initially, there was a modicum of distrust for one another and a measure of staking a claim on who was superior or more important than the other. Maintenance personnel held the position that the aircrew could not fly without their essential efforts, and the aircrew would adopt the perspective that without them, the Maintenance section would not have a reason to exist. This apparent confrontation was quite transient in nature (it took time to build trust internally between sections: one disadvantage of an ad hoc organization) but nevertheless a challenge for the detachment's chain of command to resolve.

For the Detachment Commander and Detachment Chief Warrant Officer, it was our role to permit the chain of command to function knowing that the supervisory cadre themselves were not likely to be seduced by the often inflammatory rhetoric that could be best labeled 'tribalism'. Having observed this effect in other organizations, I would propose that tribalism is likely the product of competing levels of *esprit de corps* and professional pride from affiliated sections, occupational structures, units, and even whole detachments coming into close contact with one another or being forced together (as was the case for OP APOLLO) into ad hoc organizations. It is also an individual force that we often observe as competition among like-skilled personnel; for example, competing for promotion.

At all levels of an organization, the astute leader can shape that competitiveness into enhanced team output. That being said, an organization's leadership also has a role here in preventing this social dynamic from festering and becoming counter-productive. Caution must be exercised to not permit excessive supervisory intervention to extinguish the motivational benefit that healthier competition can bring but this is a very intuitive, 'gut' personnel management activity and difficult to define. For my command role, managing this human dynamic was an organizational 'tightrope' to walk with no clear markers to lay down in defining a correct methodology to follow but is emphasized as a real force with which a leader must deal. The role of the leader in monitoring this internal social force was crucial to setting up optimal long-term working relationships within the detachment but it became even more critical when the full realization of the deployed circumstance became known to all of us, and it was something for which no one was very well prepared, including myself.

The Tactical Airlift Detachment left 8 Wing Trenton as a singular organizational entity but arrived at its deployed location as part of a larger build. Indeed, our self-sufficient operations unit was the 'second' one to arrive at the desert airfield location having already been made home to a detachment of Aurora long-range patrol aircraft primarily sourced from 14 Wing Greenwood and manned by a representative population of maintenance, aircrew, and support personnel. Having arrived a few weeks prior, they had clearly established themselves on location, shaped their work routines, and distributed their support resources and personnel in a manner that provided direct advantage to their ongoing operations. Framed against the earlier phenomenon discussed as tribalism, the arrival of the Tactical Airlift Detachment brought with it new challenges for my leadership and that of the Long-Range Patrol (LRP) Detachment Commander. We both quickly discerned that our shared partnership would have to quickly take shape in order for our respective units to co-exist in relative harmony. Without that, prevailing prejudices and distrust would threaten to dismantle much, if not all, of the levels of trust, team spirit, and potency we had each energized in our separate detachments.

Tribalism manifested itself in many ways in this physical circumstance. On many levels, it showed itself in terms of tension between professional communities (Maritime Patrol versus Tactical Air Transport); geographical disparity (8 Wing Trenton vs. 14 Wing Greenwood); organizational loyalties (Detachment versus Detachment); functional differences (Persian Gulf surveillance versus Afghanistan re-supply); and, even occupationally (support personnel from one location versus clients from another location).

The arrival of the 'invading force' was clearly disruptive for many who had already 'staked out their turf'. Regardless, the need to integrate and share our support personnel was seen by both commanders to be in our best collective interest but that organizational efficiency alone was not enough to counter the real demographic force that was being exercised. Identical to the aircrew/maintenance example described earlier in this part of the chapter, there was an immediate need for shared leadership to assure the fostering of mutual respect for the differences that naturally existed between the two detachments, their roles and their personnel. That effort had to start from the top. Both commanders understood this and engaged their supervisors to renew a broader campaign of trust and mutual respect while sustaining our own internal morale and unit output. It was now a joint leadership responsibility to ensure that: open lines of communication now existed between the two units; any accusations of animosity (real or perceived) were redressed fairly and quickly; that support personnel remained in direct contact with and still felt like they were an organic part of the 'total' air operation; and, that we ultimately became a larger 'single' operational entity: one team.

Administrative leadership of this base of operations had already been ordained by higher headquarters to the LRP Commander but to my counterpart's credit, he shared decision-making and policy creation tasks with me in an open and respectful manner. In short, the two of us became a team and the others tended to follow suit, illustrating the classic 'lead by example' model. This overall campaign of renewed unity was very much a work in progress and we were well on our way to making it work when another unforeseen organizational change took place that had immediate effect.

CHAPTER 6

In early March, a key strategic decision was made to amalgamate all of the regional theatre support elements under a single group principle. It, too, was to be collocated at the same site as the two detachments, making it the third organizational tenant. Support personnel sourced originally from within each detachment were withdrawn from those organizations and placed together with other area support units to form the National Support Unit (NSU). Their commander arrived later in March and commenced his team-building activities while the two detachments prosecuted their daily sorties into the theatre of operations. Their role was to provide unified logistics, financial, administrative, medical, and other traditional forms of support functionality enjoyed by deployed personnel.

Much like my experience, the NSU Commander had an ad hoc organization into which he had to inject his personal leadership energies so as to generate his own form of *esprit de corps* with the aim to realizing the highest potential for success. Like me, he was confronted by some new dynamics. For many personnel in this new organization, this was now the third layer of change that had been imposed (the first being the ad hoc detachment construct at each Wing location and the second being the grafting of one detachment onto another upon arrival) but the most identifiable effect of this change was the loss of direct ownership to the operation that all support personnel had enjoyed to that point in time. They immediately sensed their team colours had changed and while they were still functionally critical to the success of our respective detachments, some did not sustain the same sense of belonging in the transition: a significant destabilizing force that the NSU Commander and his supervisory cadre had to overcome.

There are two essential lessons to be drawn from these exemplary anecdotes. First, as a leader, one must be ultimately prepared for change, but more importantly, one has to make an immediate appreciation of the effect that change will have on the cooperative team-building work conducted to date. Secondly, forces like 'tribalism' and 'operational ownership' are very real and very potent. Properly managed, they can contribute directly to the success of any operation or organizational goal. Ignored, they can lead to the failure to attain an organization's goals.

Culture is Us

Not unlike the preceding tales of people interacting with one another, the relationship of individuals to changed circumstance and a foreign environment is part of the set of distractions that I will discuss next. Broken into internal and external variables, they can also be grouped into forms of behaviours or 'cultures'. For the purpose of this brief discussion, internal cultures are: reflections of the traditions and habits that shape us as trained military personnel; the social habits and value sets that we carry with us; and, the mind-sets that define our respective professional occupational structures. By contrast, external variables are the things that a leader has no control over in terms of situation or environment. They can best be described as 'culture on culture' or better related to this operational experience: 'strangers in a strange land'.

In the transition from 8 Wing to the deployed location, I had described earlier how tears were shed on departure and how families were confronted with an unprecedented period of separation but what I did not mention was the insecurity that this action brought for several of the detachment personnel. They had been removed from familiar surroundings and routine. In spite of everyone's best efforts to anticipate the degree of imposed change, they brought with them internal mechanisms to deal with this disruption – some of which were disruptive to team discipline. Foremost in this area of leadership focus, the occupational overhead of working and behaving in a security-enhanced environment was a leadership challenge. A 'home-base' mentality or what I affectionately refer to as 'holiday brain' and the complacency that accompanies such a mindset was clearly a direct threat to the organizational integrity of the unit for which I was responsible and accountable to secure. Security measures, on and off camp movement protocols, and other imposed regulation that were needed to protect the personnel and resources under my charge, were seen to be a threat to the comforts of those same personnel for whom the design was intended.

Added to this sense of encroachment was a shift-worker attitude that ran counter to the leadership intention of building one team with singular purpose. To better describe this phenomenon, the shift-worker only considers themselves 'viable' members of the team when they are working while choosing to change loyalties back to themselves when they are not 'on shift'. With local infrastructure nearby that presented opportunity for escape from these imposed working lifestyles and reinforcing the 'me' needs just described, the potential for disciplinary action was ever-present. The leadership challenge for me was seeking out means to moderate between these seductive influences. I recognized immediately that I could not personally police the behaviour of individuals around the clock so I found myself relying heavily upon two key components of my command to date: trust (through the chain of command); and, setting the example.

I ensured that standards of accountability and enforcement of discipline were universally applied. I most certainly understood that I could not allow myself to be taken in by the same temptations to which I also had full access, but the most critical aspect of my approach in setting an example was my tactic of making everyone owners of the Detachment's discipline just as I had done with support personnel and the operation output. In my mind, I believed that peer pressure was probably the best means to benignly control gross deviations in personnel behaviour but I was also not so naïve as to believe that breeches of discipline would not occur. In that regard, it was helpful to have other units on location because this presented additional objective opportunity to monitor personnel behaviour beyond the scope of direct observation. Just as the competitive forces of 'tribalism' mentioned in the previous section so repelled competing personalities in the workplace, it served me well to gain insights to the 'off-shift' behaviours of the personnel in my charge but reported upon by others seeking a means to possibly discredit them. Investigations took place and discipline was handed out but, as the leader, once the punishment had been meted out, I was equally quick to draw the offending person back into the team.

I earlier made reference to the idea of external influences over which a deployed commander may have little, if any, control. This was precisely the case that we all suffered on this operation. Deposited squarely in the middle of a Muslim nation replete with its behavioural standards and cultural expectations, pre-deployment education was critical enough but avoiding an international incident due to some breech in local laws or accepted tradition was something that troubled me greatly throughout the mission. This deployment, unlike others I had experienced in other places of the world, was operating in the absence of any formalized agreements other than the host nation's government giving its consent to our imposition on their sovereignty. The uniqueness of this arrangement made for a perplexing understanding that should anyone be in violation of local laws, there would be little or no diplomatic protection afforded them. To make matters worse, this arrangement did not allow any opportunity for me to intervene in the local judicial process, which was, to be blunt, substantially lacking the rights and freedoms to which most Canadians are accustomed.

Area legal practices were highly skewed against non-Muslims especially if the infraction had been religious in nature. My only defence in this regard was to broker a relationship with the local law enforcement agency expecting that I might be given an opportunity to influence a relevant judicial proceeding. Given the nature and duration of the punishments, the description alone provided substantial deterrent value for all Detachment personnel.

But the challenge here was not only discipline, it was also cultural behaviour. Our personnel had to respect the traditions and practices of their hosts otherwise the operation would have been jeopardized. In a leadership context, both they and I had to manage cultural behaviours imported from homes and workplaces in Canada, control these behaviours to remain within the boundaries of disciplined professionalism, and fit seamlessly into a foreign environment that was fraught with the potential for an operations disruption or even worse, a directed terrorist attack – the very thing we were sent to confront with our operational mandate. We had to continuously remind ourselves that we were not "in Kansas anymore!"

The Burden of It All

To this point, the focus has been directed at the interplay within the chain of command, the detachment personnel, and other external influences (human or situational). However, as a leader, one must never forget about looking after oneself or accepting the offers of others to help. The clichéd phrase, 'the burden of command' can only be properly interpreted after having been completely exposed to it in a situation where one's decisions can cost a human life. In the Air Force experience, manned flight is understood to be somewhat unnatural and genuinely dangerous but there are technical and procedural standards put in place along with substantial levels of individual training that prevent aviation-related mishaps from occurring more frequently. A similar perspective could be offered by specialists in the Navy or the Army but the point to be made here is that by repeated exposure we become comfortable in our respective professional environments and the hazards that accompany them.

Placing this point in the context of leadership however, we have little in the toolbox of being prepared other than escalating levels of leadership opportunity and increasing levels of responsibility for personnel and resources as we move through our careers. Consider the well-publicized case of then Major-General Roméo Dallaire in Rwanda. Little could he have understood or anticipated what that assignment would ultimately do to his life – a direct result of the burden of command; however, I am not about to compare stories here. Instead, I choose this example to give more clarity to the notion that the burden of command is indeed weighty and it will have definite, measurable effect on those who feel its full press.

Not unlike the earlier situation cited, where the need to control and deal with everything is one of the first adversarial influences a leader must overcome, a good leader must also trust his subordinates to run the operation against the same performance standards that have been personally set through individual example. This is not more dialog on the prevailing theme of trust but rather an insightful piece of advice for a young leader not to be ashamed of taking personal time. For me, my first day directly removed from my command function and centred on my own needs was six weeks after arriving. It was also the first day where I left my second-in-command largely responsible for the operation. Until that time, my working days had been typically eighteen to twenty hours long with the balance of each day left for sleep. For the time and in the moment, this was my answer to the call for responsibility but I know looking back now, it was a mistaken approach to shouldering the burden of command. My Detachment Chief Warrant Officer, for whom I shall always have the fondest regard, jolted me with the incisiveness of a statement he made to me one day. He said quite simply, "Sir, I'm not taking a day off until you do, and frankly, I'm [expletive] tired!" Many eyes are watching you. It is difficult for others to take the appropriate amounts of down time when you as the leader do not set the proper example by doing the same.

As for example setting, set a standard that is sustainable. You are under the very same social and cultural pressures but you must make extraordinary effort to not be two different people – an extrapolation of the 'shift-worker' phenomenon mentioned earlier. It is a leadership responsibility to set achievable standards of professional demeanour, ethical conduct, and self-disciplined restraint but there is still room in all of that to relax – in moderation, of course. Similarly, there are times when public displays of anger, complaint, frustration or disappointment have real application and a desired effect but heavier than normal reliance on any one of those will jade your leadership presence especially if it directly involves a subordinate. Behaviourally, do not get caught up in a kind of 'popularity syndrome' with your subordinates. Tell them what they must know, not what you believe they want to hear.

Loyalty, too, must be held most sacred. Loyalty exercised up and down the chain of command fuels *esprit de corps*. Loyalty exercised only one way, in my opinion, is exploitation. All of these perspectives are examples of the tools that I have placed in my own personal leadership toolbox but they are all also uniquely tied to my personality and style: something that will evolve in each of you as you move through your career and life experiences. Some techniques may work for other leaders and others may not.

Throughout this chapter, I have implied what the measurement of operations success could be but I have not explicitly stated what my metric was during my tour on OP APOLLO. To extend that question, how might I have measured my success in a leadership role? This last question is not an easy one but the previous question was relatively easy for me to answer. Everyone on the Tactical Airlift Detachment expected that our collective success as a deployed force was going to be measured by how much freight was carried or how our sortie execution rates compared with our coalition partners performing comparable roles. To some, these were the markers of our accomplishments as a team but for me it was relatively meaningless. Almost nine months earlier, when planning had commenced, the first time I stepped up to the podium to speak to the very first set of families to address their anxieties of knowing what it was we were going to do, the look in their eyes said it all for me: I had to bring everyone home safely. That, after all of the blistering heat, the aircraft maintenance problems, the extraordinary sortie rate we sustained, and the successes we enjoyed as an ad hoc unit put together to answer a call to action, was my metric of success: <u>we</u> brought everyone home safely. I told the Detachment that very story during a short speech on my Change of Command parade at the conclusion of my tour. Only then could I say that our job was done.

PERSONAL REFLECTIONS

I learned many things in this 'non-traditional' assignment area. I learned about my own vulnerabilities not only as a professional military officer but as a human being. I discovered new boundaries in my own ability to cope with stress or, as some would understand the phrase, I finally understood the immensity of "the burden of command". I especially discovered the enduring strength of my family and so many others who patiently supported us from Canada. I got to know otherwise anonymous Canadians from coast to coast who took the time to send cards and letters to us without purpose other than to cheer us on for the wonderful and most dangerous things we were doing in all of their names. There is no doubt that the mail deliveries were very popular just as the occasional telephone call was so dearly anticipated by each of us. We took great strength from all of this.

Morale is such a fragile thing and it demands care and attention from all commanders at all times. With other men and women focused on a common aim, we did some extraordinary things during our service to Canada. We were able to conquer most challenges but only because of the unifying and immeasurable force of the human spirit and kinship that we all shared. There was no 'checklist' to follow but by ensuring that a vital and ethical relationship existed between my Detachment Chief Warrant Officer and I, we were able to stay ahead of the troubling issues and solve most problems before they festered. Our combined loyalty and integrity in communicating with each other was visible to the entire Detachment. The team took great strength and trust from that example and morale was the by-product of that expended energy.

CONCLUSION

In this chapter, I have explored a series of what I believe to be instructive leadership markers. In sum, I learned: to understand my own performance boundaries better; to acknowledge and anticipate that others can and will make mistakes; to avoid psychological injury by singularly carrying the entire burden of command; to take time for oneself through active trust instilled in others; to set reasonable and attainable goals; and, to apply 'operation ownership' to all members of the team. Of all of these, I want to re-emphasize the last point: do not ever forget that success is shared and that one measurement of success as a deployed commander will be the total team's ability to do their jobs better than one might expect. Whenever 110 percent output happens, one has succeeded as a leader but it is because the personnel in the team 'wanted it to happen' – the leader did not make them do it, the true leader made them 'want' to do it.

To close, I cannot emphasize enough how fortunate we should all consider ourselves to live in this lush, resource-rich country. Over many years and countless travels though poverty-stricken countries, I have seen many people who regard Canadians as those who dwell 'inside' the pot of gold at the end of a desperately long rainbow. Following an event held at Rideau Hall, almost two years following my return, where I and my good friend from the LRP Detachment were awarded the Meritorious Service Medal for our leadership during OP APOLLO, to have been selflessly approached by the Afghanistan Ambassador and personally thanked for helping to rebuild 'his' country helped to put all of the effort I have attempted to describe in this paper into the proper perspective. Only now as I write do I fully comprehend the magnitude of being a truly exercised commander and the wonderful period of growth it has provided me with as a professional military officer. There can be no doubt that this has been a culminating point of emotional, spiritual, and psychological development at the height of my personal technical competency as a skilled operator. Without hesitation, I would do it all over again but knowing what I know now, I just might do it better next time.

CHAPTER 7

OPERATION HALO: A LEADERSHIP CHALLENGE

Lieutenant-Colonel Pierre St-Cyr

In the following chapter, I hope to place the scope and application of leadership during the deployment to Haiti in March 2004, as part of Operation (OP) HALO in its own context. As such, I will discuss the mechanics surrounding the preparation and deployment of members of 430 and 438 Tactical Helicopter Squadrons (Tac Hel Sqns) to deploy to Haiti for OP HALO – a mission that was completely unexpected. Specifically, I will discuss how the many challenges and issues that arose throughout the deployment, from receipt of the mission order until the return of the last member of the two Tac Hel Sqns deployed to Port-au-Prince, Haiti, affected leadership.

This chapter is not intended as a critique of every government decision concerning the situation that prevailed in Haiti at the time or the relevance of such a mission to the Haitian crisis. Leaving aside the political context, OP HALO was a success and Canadian citizens can be proud of the achievement of each member of the Canadian Forces (CF) deployed in support of the operation.

In terms of leadership, the experience gained from the mission made it possible to focus on the principles of leadership and their relevant application. For example, we discovered that a strict hierarchy modeled on an inflexible chain of command is no longer effective. Functional positions, of which there are many within an organization such as a Tac Hel Sqn, include many more advisory than command duties. This means that a more human approach must be introduced where listening plays a major role in relations between unit members and levels of command within the same unit. In the context of an operational mission such as OP HALO, then, traditional concepts of military leadership must set aside blind application of authority and instead seek to motivate, guide, develop and recognize the contribution of members to the mission and the organization.

The subject of leadership has been examined throughout human history, from the time of the Greeks to the present day and, with each re-examination, tentative definitions were put forward. The subject – leadership – encompasses a number of paradoxes, which are still contributing to our misunderstanding of it.[1] Whether learned in the classroom or acquired through experience, leadership remains an obscure concept that is attributed a little too summarily to any relation within an organization. One thing is certain: when confronted with a situation involving responsibilities to be discharged and goals to be achieved, some relationship must be established between the people in charge and the group in order to work effectively. Can leadership be applied in the same way for every situation? Is there only one "recipe" for leadership? In military operations, it is erroneous to imply that blind leadership based on the concept of authority and power and resulting in robot-like obedience makes it possible to achieve goals.* An approach based on the exchange of ideas, respect and motivation, as demonstrated during the Canadian Forces mission in Haiti (OP HALO), has proved to be an extremely effective style of leadership.

OPERATION HALO

Before examining the issue of leadership, it is necessary to provide some background on OP HALO. In early February 2004, armed conflict broke out in the city of Gonaïves and, in the following days, spread to other major cities in Haiti. The insurgents gradually gained control of a large part of northern Haiti. Despite diplomatic efforts, the armed opposition threatened to march on the capital. In the early morning hours of 29 February 2004, Mr. Aristide, the elected president, left the country. A few hours later, Boniface Alexandre, President of the Supreme Court, was sworn in as Acting President, in accordance with the constitutional provisions governing succession. On 29 February, Haiti's permanent representative at the United Nations (UN) submitted a request for

---

* Editor's Note: Updated CF leadership doctrine was promulgated in March 2005. As such, effective military leadership is defined as "directing, motivating and enabling others to accomplish the mission professionally and ethically, while developing or improving capabilities that contribute to mission success." See Canada, *Leadership in the Canadian Forces Conceptual Foundations* (Kingston, ON: DND, 2005). Moreover, the concept of leadership should not be confused with command, which is defined as "the authority vested in an individual of the armed forces for the direction, co-ordination, and control of military forces."

assistance from the acting president authorizing international peacekeepers to enter Haiti. Further to this request, the UN Security Council passed Resolution 1529 authorizing the deployment of the Multinational Interim Force (MIF) and declared that it was prepared to create a United Nations stabilization force to facilitate the continuation of a peaceful, constitutional political process and to maintain conditions of security and stability. In accordance with Resolution 1529, the MIF immediately commenced deployment to Haiti.

The CF deployed a Task Force to Haiti (TFH) of nearly 500 soldiers and six CH-146 *Griffon* helicopters to help the MIF restore stability in the country, which had descended into anarchy. The CF contribution was named OP HALO. TFH was deployed in early March 2004 as part of the United Nations MIF. Established in late February, the MIF was mandated to remain in Haiti for 90 days to contribute to a secure and stable environment in the country, to facilitate the delivery of relief aid to those in need, to help the Haitian police and the Haitian Coast Guard maintain law and order, and to protect human rights. It is important to note that TFH was deployed very quickly and had to endure extremely primitive conditions, comparable to other CF operations such as OP TOUCAN (East Timor) and OP DELIVERANCE (Somalia). Furthermore, the Canadian chief of staff (COS) of the mission and the senior leadership of MIF acknowledged and praised the unconditional commitment, utmost professionalism and boundless dedication of the members involved in OP HALO, which, in their view, ensured the success of the operation and secured the reputation of the CF.

CHAPTER VII MANDATE

It is also important to understand the type the mission that was required due to the situation. The fact that this was a Chapter VII mission (under the UN Charter), distinguished it from previous peacekeeping missions, normally covered by the more benign Chapter VI mandate. The Security Council determined that there was in fact a threat to the peace and a very great risk that acts of aggression would be committed against MIF personnel and facilities. Therefore, they decided to deploy the MIF under Chapter VII of the United Nations Charter, entitling it to "Action with Respect to Threats to the Peace, Breaches of the Peace, and Acts of Aggression". The chapter defines the conditions under which the Security

Council may authorize the use of force to "give effect to its decisions" in case of any threat to the peace, breach of the peace, or act of aggression.

As a result, it clearly understood that the MIF was deploying in a theatre of operations in which armed groups—in this case, rebels, the Chimères and former members of Haiti's military—had destabilized the Haitian population. The Security Council's mandate accordingly authorized MIF members to adopt a robust posture with weaponry that created a deterrent effect. The rules of engagement (ROE) governing the use of force by both the UN and the CF were consolidated, allowing members of the MIF and TFH, in certain circumstances, to use "all the means necessary" to protect civilians in their immediate vicinity and prevent violence against UN personnel. In addition, the MIF protection concept required CF members to wear bulletproof vests and to be in possession of their steel helmets and personal weapons at all times. It was, therefore, clear beyond a doubt that the MIF was facing armed antagonists.

THE TEAM

As has already been noted, the helicopter detachment (Hel Det) personnel came from 430 Tac Hel Sqn Valcartier (84 members) and 438 Tac Hel Sqn St-Hubert (6 members). The 430 Sqn is unique in that it is the only "Francophone" helicopter squadron in the CF. A "Latin" culture has developed during its 62 years of existence, especially over its last 33 years in the Quebec City area. The members of 430 Sqn form an informal group within tactical aviation (Tac Avn). There is an obvious bond between squadron members, expressed by the need to resist the organisation (i.e. the CF). Emotional ties, which are closer between squadron members than with other members of the Tac Avn community, have resulted in greater cohesion among them. These members demonstrate solidarity, provide mutual assistance and are constantly exchanging items and services. In times of danger, squadron members draw together to establish a protection and support system, thus ensuring harmony among them. The usual taboos of the organization no longer faze squadron members or prevent them from achieving their goals.

The behaviour of 430 Sqn members corresponds to the description of an informal group, as may be found in literature; members of the unit cultivating strong interactions by a combination of the time devoted to them, the emotional intensity, the closeness and the exchange of services, or by the fact that these people interact more among themselves than with other members of the organization.[2] Thus, 430 Sqn became the nucleus of the Hel Det. Of course, the homogeneity sought for this deployment already existed within this new team. For these people, participation in the mission allowed them to be away from the main organization, creating a certain sense of independence. On the other hand, a special approach in terms of leadership was required of the Hel Det commander in order to provide a framework for each member's actions.

EVOLUTION OF LEADERSHIP WITHIN THE CF

Before describing the type of leadership that prevailed during OP HALO, it is necessary to identify the context in which leadership is acquired within the CF. Many authors are in agreement that we become leaders and that leadership is developed as we acquire experience and maturity.[3] Leadership can thus be defined as behaviour.* This calls into question the rationale of any so-called leadership school in cases where such behaviour is only acquired from personal experience, rather than learned from academic studies. The concept of learning the rudiments of leadership by studying theory has caused a number of problems within the CF. The CF used training programs to imbue members with leadership qualities. Yet, the CF made little or no distinction between a manager and a leader and this approach has contributed to the development of the image of the Canadian soldier as more civil servant in uniform than soldier. In his book, *The Administration of Defence Policy in Canada 1947-1985*, Douglas Bland reiterates this duality between the CF's administration and operations, showing that the CF trained more managers than leaders during that period.[4] The CF have acknowledged their error, especially following the events in Somalia, in awarding leadership positions to people who had received theoretical training in the classroom but had very little experience in the field.[5]

In October 1997, the Minister of National Defence (MND) made a number of statements in response to concerns expressed by the Somalia Inquiry about leadership standards within his department. First and foremost, he decided that the selection, professional training, professional

---

* Editor's Note : See footnote on leadership and its CF doctrinally accepted definition on page 138.

development and officer evaluations would be based on formal criteria. He also tasked the Chief of the Defence Staff (CDS) with developing accountability criteria for persons in positions of leadership within the CF. The Minister decided that the criteria defined by the Somalia Inquiry would be used for personnel screening, selection, promotion and supervision, delivery of basic military training and continuing development, the administration of discipline, the chain of command, operational preparedness and mission planning.[6]

The pre-1997 events, with their leadership failures, led to the awareness that leadership scholars such as Bernard Bass and Bruce Avolio warned about. Namely, that leadership must adapt to the changes that an organization is going through. Leadership based on antiquated concepts of authority or power no longer satisfied the needs of the organization in light of the changes experienced by the CF. The Somalia Commission noted:

> Our findings regarding current patterns of leadership in the military may suggest that many top-level officers might have been promoted on the basis of their transactional abilities to work within the system. However, the military is undergoing some fundamental changes, which may result in a different type of leader emerging at the top. We may see more Norman Schwarzkopfs who display all of the transformational factors and less of those generals who know how to 'work the system' transactionally.[7]

As a result of the Somalia Commission of Inquiry, the CF had to review their attitude toward leadership. It became critical to follow the recommendations[8] and display and demonstrate the transformational factors of leadership, in order to prevent the perception by the troops that "leaders take advantage of every opportunity to build their empires and follow their personal agendas."[9] The key element in applying this type of leadership is to establish a relationship of trust in order to overcome this perception. Once that trust has been established, the leader's influence on the team becomes a positive factor. To achieve it, the transformational leader must follow their own characteristic process, since one cannot stand before a team and expect to obtain concrete results if a relationship of trust and authority has not been clearly established.

Such a relationship leading to effective group leadership gradually developed between the Hel Det members and myself. By following this process, I was able to guide and inspire my subordinates and to be accepted by them. I believed, and still believe, that this was the type of leadership that the Department of National Defence (DND) was looking for.

MODEL OF A TRANSFORMATIONAL LEADER

Leadership literature describes three models of a transformational leader. The first is the transactional operator pursuing a personal agenda without wasting time on the concerns or well-being of others.[10] Conversely, the second, is the team player with the ability to motivate everyone while obtaining respect and trust from his team-mates. However, this type of leader is heavily influenced by the prospect of losing popularity among the personnel. In a sense, the entire team functions as a leader, while the leader merely ratifies the team's decisions. This close relationship or symbiosis is largely based on the benefit to the team alone, leaving little or no autonomy for the individual development of its members.[11] The third model is the "self-defining leader," as exemplified in my behaviour as Hel Det Commander. While considering the preservation of team cohesion based on mutual respect among team members, I nevertheless maintained a less populist attitude, in order to achieve the goals of the mission. One method I frequently used was to delegate duties to team members in order to develop their potential and responsibilities. Thus, I divided up all the duties of the organization, team and individuals, and ensured that they were distributed at the appropriate time and in such a way as to maximize each member's performance and development.

The transformational leader adopts a style that encourages the development of a strong sense of responsibility and initiative among his subordinates. To achieve this, he establishes relationships with his people and achieves communication with them. Lieutenant-General W. E. Ulmer Jr. of the United States (US) Army, describes the concept of transformational leadership in the following terms, "transformational leadership, by the enlightened use of inspiration, communication, and understanding of human behaviour, can motivate subordinates to achieve more than could ordinarily be expected."[12]

A similar perception exists within other public organizations, including the Public Service of Canada. According to them, the transformational leader "infuses a vision and a mission, inspires and stimulates staff by transferring to them his pride in work completed, builds confidence, maintains high levels of conduct and performance, encourages critical thinking and problem-solving, gives the individual attention that each one needs and treats his people with respect. He must possess a keen sense of self-understanding and personal identity."[13] Both definitions capture the essence of transformational leadership, communication and the relationship between group members and the leader.

CONTEXT OF OP HALO

In order to establish communication and a good relationship with one's subordinates, one must, above all, analyse the situational context facing the leader and his group and identify any elements of change from the normal state of affairs. Clearly, a major distinguishing factor in conducting an operational mission such as OP HALO is the dramatic departure from the normal garrison routine. This requires members to adapt immediately in order to launch the operation in an efficient and effective manner. The two critical factors are comprehension and speed, as they will dictate how leadership is administered. In the case of OP HALO, Hel Det members were faced with the following organizational changes:

    a. Acceptance that they were operating in a non-permissive environment, owing to the devastation, lack of services and inherent risks in unsanitary conditions;

    b. Rigidity, given the many security regulations and strict operational procedures;

    c. Discomfort, where crowding and lack of privacy were ubiquitous;

    d. Risk, owing to the continuing conflict situation with the Chimères; and

    e. The unknown, owing to unfamiliarity with the culture, manners and customs, as well as the Creole language.

As mentioned earlier, a firm foundation within the Hel Det would ensure speedy adaptation. That foundation was to be achieved by securing greater cohesion. The first challenge for the leadership was to develop team spirit. This raised the issue of socialization within the team and with other members of the contingent and mitigation of the effects of distance between Hel Det members and their families. To achieve this, favourable conditions had to be created for producing high levels of efficiency and satisfaction and for encouraging teamwork.

While still focusing on the mission's objectives and actual conduct, projects were introduced within the team to quickly develop a sense of achievement and pride. The fact that the mission was carried out in primitive conditions demanded efforts on a daily basis to improve living conditions. In such a context, there was no room for individualism – everyone had to pitch in. The projects included a school renovation,[14] daily participation in Civil-Military Cooperation (CIMIC) projects, construction of wooden floors and installation of additional tarps to keep out water during torrential downpours. Exploration and reconnaissance trips in the new territory; working at reducing Haiti's inherent risks and dangers; decorating the premises with items identifying the group; establishing and validating Hel Det flight procedures; organizing work areas; allocating specific responsibilities to each Hel Det member by establishing a work breakdown structure (WBS); and free work management planning were only a few of a long list of projects that allowed them to develop a very strong team spirit.

Thus, conditions were created to enable Hel Det members to gain experience against this new operational background, to examine and take greater responsibility for their actions and to derive satisfaction from their actions through self-reinforcement.[15] Anecdotally, this supports research that suggests that transformational leadership has a positive effect on innovation.[16] By stimulating my team members in this way, I was, in fact, appealing to their sense of innovation and satisfying their need for achievement. The team members carried out projects that were practical for them, for the entire team and for the Haitian people and that will definitely survive after their departure.

These projects helped satisfy the need for identification and they had the effect of forging strong bonds between Hel Det members, thus boosting team spirit. The cohesiveness and homogeneity forged among the team members were crucial ingredients in the recipe for working effectively and achieving goals.

In addition, guided by my own value system and personal standards, I promoted an environment in which members of my team were encouraged to:

    a. approach problems and issues creatively;

    b. become personally involved in problem resolution;

    c. analyse risks;

    d. where necessary, ignoring organizational inflexibility; and

    e. take appropriate initiatives.

On this issue, the leadership literature clearly describes the positive results of the difference between organizational and individual values. When clearly expressed in terms of the necessary loyalty toward the organization, personal values contribute to the development and achievement of a new level of organizational maturity. By having people who are aware of their development and capable of fulfilling their potential by agreeing to take on and successfully complete the various tasks and responsibilities assigned to them, the organization can only become greater.[17]

The effect on members of an organization when they discover that their personal values may influence and even change organizational behaviour can only be to raise their awareness. This change may produce a reinforcement effect, thereby increasing the members' self-esteem, while establishing team spirit. By being aware of this dynamic, we can expect continued improvement in the quality of leaders within the CF.

During the deployment, I found that if a situation requiring a team response and drawing on the potential of all Hel Det members arose and was not controlled, it could seriously affect, and even compromise, the future of the mission. Haiti is located in a geographic region subject to tropical storms, which in June directly threatened the island of Hispaniola, thus putting the Canadian contingent at risk. The primitive facilities, which left personnel and equipment almost totally exposed, were not built to withstand strong winds and heavy rainfall. The emergency situation demanded a rapid response to develop protective measures. Structures had to be reinforced, equipment, vehicles and personnel protected and aircraft evacuated.

Aircraft posed a number of problems because the locations considered for their evacuation had to be out of the storm path, yet close enough that aircraft could be returned quickly to support efforts similar to those during the flooding of Mapou and Fond Verrette. It was becoming clear that the aircraft should be evacuated to other locations, such as the Dominican Republic, Guantanamo Bay (Cuba), Nassau (Bahamas) or Florida (United States). A feasibility study of each location under consideration was conducted. It soon became apparent that it was becoming impossible to deploy the aircraft and we were forced to keep them in Port-au-Prince. Solutions were proposed for the immediate situation, such as chaining the aircraft to the ground, surrounding them with heavy vehicles to screen them and covering them with tarps.

Tents were secured by cargo straps and stakes, all equipment was placed in sea containers and measures were taken to accommodate personnel in buildings at the Port-au-Prince airport. The emergency situation demanded immediate results, a direct contribution by Hel Det members, awareness of each member's potential to find a solution and maintenance of team spirit. The situation also helped establish the Hel Det as an intelligent organization that could both communicate and learn, since the entire Canadian contingent benefited from the protective measures introduced by us and witnessed a clear demonstration of our dynamism and potential.

The results would have been different if the chain of command had been maintained, since decision-making would have been slower, resulting in delays that might have made the situation more dangerous. An analysis of the situation demanded immediate action and only by boldly omitting certain levels of decision-making could an action plan be put in place in time, resulting in a dichotomy between my values and those of the organization. This by no means represents disloyalty on my part, as I remain convinced that my action served the purpose of the mission. Through this interaction, Hel Det members achieved personal and mutual empowerment, while demonstrating interpersonal commitment and cooperation.

The situation demanded a leadership that was always in a listening mode in order to analyse the requirements of the situation quickly. It was the appropriate time to demonstrate mutual trust between my team and myself. My people were professionals who, because of their technical experience, provided advice and offered potential solutions. It was the appropriate time to demonstrate that I trusted them; in return, they always supported my decisions. I never had to exert my authority to convince them, as my decisions were based on their advice and ideas. This trust resulted in the development of a highly effective synergy and the rapid implementation of protective measures.

Another important aspect of my leadership style was that I never presented the traditional concept of an old-fashioned personality in which the individual considers himself the centre of the world and those around him a negligible quality. "Remember that a group dominated to too great an extent by the leader will never succeed in overcoming the limitations imposed by the leader's talent."[18] That kind of centric attitude limits organizational development. My concept of command is entirely based on the concept of human nature and the nature of interpersonal relations.[19] It would have been strange not to apply that concept of harmonious relations, listening and respect to the Hel Det team when the primary mission of the Canadian contingent was to assist in bringing social stability in Haiti by encouraging harmony and mutual respect among the various groups present.

A transformational leader must focus on the goals of the mission while considering how to satisfy the needs of the group, i.e., a definition of the structure, goals, methods, resources, type of supervision and communications of the group. By assuming responsibility for these basic needs, the leader places himself in a position of influence. The extent of that influence becomes the key to assessing the leader's impact on his group. First and foremost, the leader must listen to his group and become a part of it, if he wishes to exert a positive impact and achieve the goals set, rather than simply to exercise power. This has a significant impact on how he controls the group, especially in an operational context, such as the Haitian theatre.

FUNCTIONS OF A TRANSFORMATIONAL LEADER

A transformational leader must perform various functions to strengthen his credibility and demonstrate his leadership to his team. As a commander, I performed four important functions: coaching, management skill, technical skill and humour.[20]

Coaching

Coaching is by far the most important function of a transformational leader. I demonstrated this function by motivating and encouraging my people to become involved and take an active role in the mission. First and foremost, the mission and its concept had to arouse a sense of duty. I therefore presented a clear, precise explanation of why we were in Haiti and a definition of each person's role to all Hel Det members in order to establish the relationship between them and myself. I established personal contact with each member of the Hel Det and maintained that contact throughout the mission. I made a point of listening and showed a special interest in each member of my team; I knew their first names and their spouse's names and I marked important dates, including their birthdays.

Because they understood the goal of the mission and my own vision, it was clear to the team members that I was showing consideration for them and their families and they were fully aware of my expectations of them. My respect for the dignity of individuals made it easier for them to describe their qualities, to understand how much room they had to manoeuvre and to operate with assurance and self-confidence. It is essential that the leader establish and maintain excellent relations with the members of his team.[21] Then he can help them react in a manner that does justice to each member's own dignity and can give each member sufficient space in which to realize his potential.

The coaching demonstrated by the Hel Det leadership made it possible to apply the principle of delegation, not to deflect a number of duties but to allow other members to develop their potential. The golden rule was applied to delegation. The leadership never used delegation to avoid certain obligations or purely for the purpose of transferring authority to another individual. In every case of delegation, the Hel Det leadership remained available as a reference for the person assigned to the task, especially in order to provide guidance in how to successfully complete the task assigned. In this way, the leadership assumed the role of teacher for the subordinates and contributed to their development while improving their skills and potential. The use of delegation had the effect of encouraging each member to become involved in the mission. However, I retained overall responsibility for the achievements, actions, success and failure of the mission at all times.[22]

My interaction with my team through coaching is also explained in the leadership literature by the fact that it involved more than a simple exchange of ideas or a good understanding. In fact, my coaching function was based on an attitude or conduct of a higher order, involving four specific interactions, commonly identified as "The Four I's" and described in the research conducted by Bass and Avolio:[23]

Idealized Influence: I made sure that my own behaviour, and that of other members of the Hel Det leadership, served as a model for the other team members. In order to achieve this, I had to be honest and ensure that I had a value system in place on the basis of which all my interactions and actions could be judged. At no time could I reveal any discrepancy in how I appeared, in good times or bad, because it would have affected my personal identity. Integrity is less concerned with our actions than with our personal identity, which, in turn determines our actions. Thus, integrity makes it possible to predetermine a leader's attitude regardless of the circumstances. Because he is aware of the effect of his attitude, a leader is also aware that his promises carry enormous power, the power to strengthen and inspire people. The organization is in serious trouble when a leader's promises become inconsistent and are no longer respected, as the group's expectations are no longer fulfilled, which affects the leader's integrity and honesty.

Of course, this concept of integrity is inevitably tied to the concept of credibility. Integrity helps the leader become not only competent but also consistent, which strengthens his credibility. "The more credible you are, the more people trust you and, thus, grant you the privilege of influencing their lives. The less credible you are, the less people trust you and the sooner you lose your position of influence."[24]

Some people react to their position of responsibility by ensuring that their actions and deeds are consistent with organizational needs, thus rejecting any initiative that could be viewed negatively by the leadership of an organization. To a certain extent, this casts doubt on a leader's degree of integrity, if his value system is impeded by the organization's reactions to the issues. As Hel Det commander, I had to begin by setting high personal achievement criteria in order to set an example for members of my team and define my character accordingly. Only then did I set high criteria for members of the Hel Det team. I had to assess the extent to which my personal values differed from those of the CF in order to maintain the focus of the mission, to provide a positive influence on the group by constantly demonstrating integrity and to demonstrate my loyalty to the organization. In order to do so, there must be a firm awareness of the individual's values and, if there is any divergence, an acceptance of the organization's response.

Team members appreciate this attitude because it is anchored in the leader's determination and demonstrates his belief in his personal values, thereby increasing his integrity and credibility in the eyes of the group. The sharing of the same values between other group members and myself made it possible to build a relationship of trust between everyone, thus guaranteeing unity within the team. I have faced a dichotomy between my own values and those of the organization on a number of occasions, particularly in the selection of members of the work team, when the organization favoured different individuals than I had chosen. The fact that I did not change my mind was noted by the team members. As a result, I was able to influence the group and achieve the desired operational effectiveness.

Inspirational Motivation: Another aspect of leadership behaviour was the constant guidance that I offered the work team. At the height of the mission, especially during the flooding at Mapou and Fond Verrette (May 2004), the team strengthened its bonds by demonstrating its determination to accomplish with the utmost professionalism each mission assigned to it. The leadership of the Hel Det team ensured that the goal of the mission was upheld and that each member's work was acknowledged as being important. Tying the professionalism of the team members to the success of the mission had a motivational effect. My involvement, and that of key members of my team, also became an inspiration to others. Participation of the leadership in preparing for the missions and supervising the decisions of junior members, by giving advice, had a positive motivating effect on everyone. Furthermore, adopting an open-door policy provided opportunities for discussion, communication and exchange of views on the decisions made.

Another motivational aspect was the use of delegation, as mentioned in the earlier discussion of the coaching function. Delegation proved to be an effective tool in the development of team members in order to increase their ability to believe in their potential and, as a result of the trust they put in me, their motivation.

Offering recognition, by giving judicious credit for the work done by the team, was also a motivating factor. It encouraged each member to realize his full potential and make the greatest contribution possible. Recognition increased individual self-confidence and trust in other members of the Hel Det team, while strengthening relations among them and increasing their sense of belonging. "Consistency in values is important in a context in which recognition is offered."[25] This reinforces the leader's integrity and credibility by demonstrating fairness and impartiality when the team's successes are recognized. Accordingly, the leadership attributed the success of difficult missions to those who had planned, executed and supported the missions. This finding is supported by Bass and Avolio, whose research concluded that "transformational leaders are perceived as effective leaders when their attitude inspires people to explore and innovate, while offering support when mistakes are made."[26]

Intellectual Stimulation: Being a firm believer in initiative, I encouraged the members of my team to adopt new approaches to the goals set. The conditions prevailing in this particular theatre of operations meant that I had to appeal to their sense of resourcefulness. While complying with safety and efficiency criteria, I encouraged the members to revise their approaches to work methods and report on the efficiency achieved. In the aircraft maintenance section, this attitude made it possible to complete aircraft inspections in much shorter times than had been the case in Canada. The flight crew developed new flight procedures, thereby increasing air safety and operational efficiency.

Individualized Consideration: Individualized consideration is the cornerstone of a team's performance. It explains how Hel Det members were able to contribute fully at higher levels of performance, having had the opportunity to develop their potential by participating as individuals or team members.[27] The involvement of team members in day-to-day operations enabled them to take responsibility for their own development. (This way, they became players rather than remaining on the sidelines as spectators.)

I considered my team members in terms of what they were outside the military: family men and women who run a family business, and deal with mortgages, personal loans, family finances, their children's education, family well-being and marital fidelity. Therefore, it can clearly be assumed that they understand the concept of responsibility. This principle of family manager and head of household allowed the Hel Det members to realize their potential, knowing that their potential was a reality that I could not ignore. This degree of trust placed in them made it possible to quantify their personal contribution to the team's achievements, while doing justice to their dignity. This recognition led to much more open, honest communication between the leadership and members of the Hel Det, in recognition of the human element of the leader/subordinate relationship, which has been rarely noticed under previous commanders.

Management

Due to the fact that the mission was unexpected, this deployment required a Hel Det leadership with an aptitude for management. For the sake of brevity I will refrain from listing every problem that arose each day, demanding resources, time and, especially, relationship management. Nonetheless, I demonstrated my management skills by setting clear, achievable goals, by displaying decisiveness and by assessing the risks to which each member of my team was exposed. This was especially noticeable in the tension arising during patrols in Port-au-Prince and flights over inhospitable regions.

Achieving a balance between protecting crews with the addition of armoured plating and the installation of door guns and the fuel requirement posed a dilemma. A low margin of power was available to the CH-146, making it impossible to completely protect the aircraft and still have the necessary capability to cover the area of operations. After conducting an in-depth analysis of the situation and considering the information received from intelligence services, I decided to keep the armoured seats as the only protective measure, which gave us enough fuel for two hours of flying time. It was a question of balance. We tried to ensure that we had the best protection for personnel and equipment while supporting the mission's goals.

Another significant factor requiring strict management was the pervasive fatigue. The primitive conditions to which we were subjected made it impossible for people to recover from their daily fatigue. Activities were always performed and duties were assigned taking into account the fatigue of team members. This required an awareness of how tired each member was, and an ability to listen and then identify the right resource for the task. Fatigue management made it necessary to undertake initiatives such as installing a pool and an air-conditioned tent, as well as monitoring the work schedule.

Stimulating the team under these conditions was a significant challenge for the Hel Det leadership. The approach selected involved examining each section's work from its own perspective, and taking unit and organizational requirements into account. Thus, the team was given responsibility for upholding my vision and pursuing consistency in its actions. This situation allowed me to encourage team members to develop their skills and potential in the performance of their duties in terms of activity planning, aircraft maintenance and flights. Motivation was a decisive factor in managing the group. It was critical for team members to be able to balance their desire for survival under difficult conditions with the interests of the group, in order to maintain a high degree of professionalism and efficiency.

The results speak for themselves. The team achieved much more than was originally demanded of them and greatly exceeded expectations. Hel Det clocked over 1,500 flying hours in 750 missions during OP HALO. Only two missions were cancelled, due to severe weather conditions. No losses were sustained during the difficult, complex, high-risk missions, all of which were deemed highly effective by all users.

Technical Skill

Technical skill plays a critical role in the group's recognition of leadership and is one way for the Hel Det leadership to demonstrate its ability to carry out missions and thus to assume and share risks with all other members of the unit. Since this was a new theatre of operations presenting considerable uncertainty for all team members, it was critical for the leadership to be the first to manage and reconnoitre this new area and to quickly familiarize itself with all the imperatives related to the missions. Thus, by setting an example in performing the first missions, being the first to expose myself to risk, demonstrating the necessary technique for mission requirements, and displaying an understanding of the maintenance problems faced by the mechanics, I established my technical skill in the eyes of the team members. I did not confine myself to managing others and to strict delegation. By playing an active role, I demonstrated my technical skills and was able to obtain greater credibility in the eyes of the team members.

By applying these three factors—coaching, management skill and technical skill—I ensured that each and every member of the team was accountable for their actions and maintained their interest and determination with respect to the mission.

Humour

Humour was a unifying factor that allowed me to approach difficult situations with greater ease and project a more human dimension for unit members, while considerably reducing tension. Humour could be expressed directly in an amusing way to reduce barriers between people. It encouraged creativity while improving productivity. It contributed in large measure to keeping morale high and reducing the impact of primitive conditions. To provide an opportunity for humour, the daily weather report included funny photographs of unit members. The weather report soon became the single daily attraction that no one wanted to miss.

THE RECIPE: FIVE LEVELS

In order to exercise productive, effective transformational leadership, I based my actions on a logical concept of progressive relations between the leader and his group. This so-called "five-level"[28] concept is based on a series of steps, each of which adds a new level to the relationship between the parties.

Level 1: The Position

At level 1, the leader's presence in his group is based on the power conferred by his rank. Initially, this is effective, as the concept of power establishes control over the group. Later, however, personnel will limit their obedience to this official authority. Continuation of the situation will soon have a negative effect on group morale. The leader must try to establish his authority over the group and go beyond the notion of power to do so. Power is an asymmetrical relationship based, neither more nor less, on strength, in which the party holding the higher position often possesses all the potential for reward and punishment, while the subordinate has only very weak control over the system of rewards and punishment. When people are under the influence of any power, stresses soon emerge, producing dissatisfaction and disagreement, and sometimes even confrontation.[29] According to scholar Chantale Leclerc, power is exercised:

> ...when there is no recourse to dialogue in order to impose one's will. It is used as a shield or defence mechanism against the anxiety, uncertainty and emotional intensity of human relationships. We believe that relationships based on power interfere with the quality of the relationship; they command attitudes that are not conducive to openness to oneself and others.[30]

A leader must quickly climb to the second level in order to achieve a relationship of authority with members of his group. If he does not, his influence will never extend beyond the limitations of his job description. I spent very little time at level 1. In fact, since the group had to be prepared for the mission very quickly, I was forced to delegate and distribute work among my subordinates. This way, I established bonds between members of the group by clarifying my responsibilities toward them, identifying the focus to be maintained and offering to lend an ear and give my full attention to any new ideas or initiatives.

Level 2: Permission

The leader will know he has reached this level when he realizes that his people are working for him without being forced to do so. I devoted my time, energy and attention to their individual needs and desires. I could not continue to have an attitude characterized by power if I was to achieve stability. I had to pay special attention to each member of my team in order for him or her to acknowledge that I was there to listen and in a position to improve each member's situation. By displaying affection for my subordinates, inspiring them to seek success, amending procedures that were often unwieldy and unnecessary for the benefit of my people, and ensuring that each and every one of them participated in Hel Det activities, I established the human contact that is necessary in a leader/subordinate relationship.

During this period, it was clear that my people were less concerned about establishing how much I knew than about establishing how much I cared about them. John Maxwell offers the following incisive summary: "leadership is developed through human relationships and not a mass of regulations."[31] By devoting over 80 percent of my time to my team in my initial months in the position, I established communication enabling me to identify each member's potential and establish each member's willingness to accept a variety of responsibilities.

Level 3: Production

Efficient production by a group is the direct result of the quality of the leader/subordinate relationship established at levels 1 and 2. Establishing training criteria and achieving each phase thereof, in order to obtain an effective, professional detachment, are the cornerstones of level 3. The goal of this level is the result. At this stage, members are happy to work as a team in order to achieve results.

Time constraints, weather conditions and primitive living conditions had an enormous effect on training and on the first weeks of activity after Hel Det was deployed to Haiti. However, the determination and pride of each member of the Hel Det unified the group and allowed us to achieve the expected result: to be ready and operational. However, one thing stood out: the majority of Hel Det members undertook, on their own initiative, to execute projects even when those projects required skills outside of their sphere of expertise. For example, the quality of life of the Hel Det people was improved considerably by laying wood flooring, installing protective tarps, putting up decorations, revising flight procedures and undertaking various initiatives to improve the work areas. We discovered that many hands achieve significant results even without the necessary qualifications.

During that phase, I was an agent of change. By communicating my vision while retaining a critical approach to my own results and encouraging my people, I incited Hel Det members to perform consistently and effectively. The production phase continued throughout the deployment. The only flight cancellations were on account of weather conditions hazardous to safe operational management. At all times, both crews and aircraft were capable of providing the expected air support. Such results would not have been possible without the strong desire of Hel Det members to deliver what was expected of them with enthusiasm and professionalism.

Level 4: Evolving as Human Beings

At level 4, the leader focuses on his commitment to developing his people in order to pursue growth and change within the organization. At this level, the relationship shows even more signs of closeness between the leader and his subordinates. The Commanding Officer expresses his belief in the competence and potential of his people, while guiding them toward personal development and career advancement. The ultimate, subtle goal that every leader should have in mind at this stage is to enable his personnel to be trained to a level of qualification such that, one day, that same leader would accept direction from members who are currently his subordinates. Such a philosophy would ensure constant improvement in any organization.

I allowed this type of relationship to develop within my team when the Hel Det deputy took charge of the Hel Det while the advance party was deployed, in my absence. Until the party's return, the deputy did a remarkable job of leading the members of the Hel Det, thus demonstrating the appropriateness of the delegation and attesting to the development of his abilities. This situation showed that the deputy had achieved a degree of development sufficient to assume the responsibilities of a higher position. This would have been impossible without my desire for the advancement of my subordinates.

Level 5: Personality

An individual's personality takes shape over time as he demonstrates his values and adopts a particular behaviour. Few people achieve a situation in which their personality has taken shape and is perceived as that of a leader. This occurs mainly as a result of the person's exposure to specific groups and situations. The length of their exposure leads us to the conclusion that they have a leader's personality because their qualities never change. In order to achieve this, such people will have devoted time and resources to developing their subordinates and their organization. In my own case, I make no claim as to whether I have achieved Level 5.

The following table summarizes the five levels of leadership:

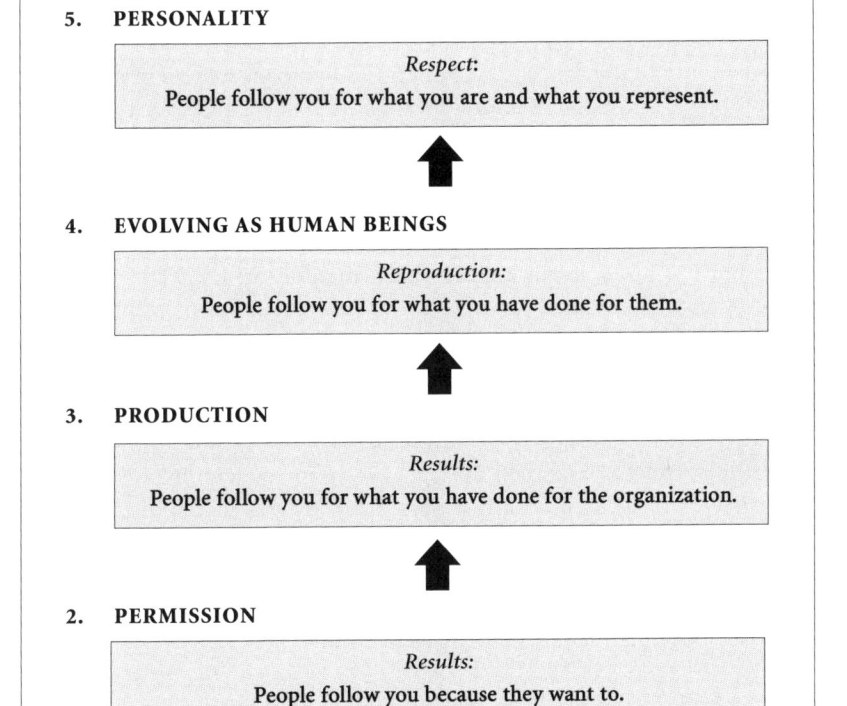

5. PERSONALITY
Respect:
People follow you for what you are and what you represent.

⬆

4. EVOLVING AS HUMAN BEINGS
Reproduction:
People follow you for what you have done for them.

⬆

3. PRODUCTION
Results:
People follow you for what you have done for the organization.

⬆

2. PERMISSION
Results:
People follow you because they want to.

⬆

1. POSITION
Rights:
People follow you because they are forced to.

CONCLUSION

Leadership has always been the ability to influence people to do a job or adopt a particular behaviour. Based strictly on a concept of giving direction, clearly power is the catalyst in the relationship between members of a group and its leaders. This implies that the authority sought by the leader is based on his ability to reward and enforce, not on his competence and credibility.

I have attempted to provide a comparison of the concepts of management and leadership, through the context of OP HALO, in order to establish the obvious distinctions between them. While management is an operational function focusing on processes and systems, leadership focuses essentially on interaction with people. Although the two concepts are not synonymous, they are complementary and must be absorbed by the team leader to ensure the success of the mission. It is also important to establish from the start of the mission, and especially in the team-building phase, individual and organizational values, in order to avoid any ambiguity concerning how the work is to be performed or how the goals are to be achieved, thus harmonizing each member's needs in terms of achievement. This distinction in terms of values requires the commanding officer to apply the concepts of leadership to a much greater extent than those of management.

From leadership literature, including a body of research by a number of authors and the 5-level model, I have tried to demonstrate that the relevance of leadership within a work team resides in the qualities of a transformational leader, who demonstrates conviction in his own abilities and a belief in those of his people, but especially in the value of his personnel. This attitude leads him to foster encouragement, initiative and, because of the high level of trust, delegation.

I have also tried to establish that within an individual project or work team, a team leader who is assigned a complex task must, first and foremost, set up a functional team and allow it to grow in terms of the responsibility assigned each member. The experience of the Hel Det in Haiti has demonstrated that the transformational leadership that I exercised was especially significant in that the dynamic made it possible to level the hierarchy and replace it with greater interaction between the different ranks.

OP HALO placed significant demands on the leadership of the Hel Det, namely the qualities of courage, boldness and a demonstration of loyalty to both troops and military authorities. Those qualities in turn produced the credibility of, and trust shown by Hel Det members to, the unit's leadership. In sum, I have concluded that persons occupying positions of command must possess the qualities of a transformational leader, i.e., a keen sense of self-understanding and personal identity in order to manage the human resources of his team appropriately and thus avoid smothering their potential. The relevance of leadership within this kind of team thus stems from a combination of authority and management, in which the demonstration of enormous credibility and unfailing integrity are essential ingredients. Leadership is much more than a job title; it is a conviction.

ENDNOTES

1   Danielle Desbiens and Pierre Comtois, *La formation au leadership: un paradoxe,* in *Leadership et pouvoir, équipes et groupe* (Cap Rouge : éditions Presses Inter Universitaires, 1998), 8.

2   Luc Brunet, André Savoie and Michel Rousson, *Les groupes informels dans les organisations: de réjouissantes données empiriques,* in *Leadership et pouvoir, équipes et groupes* (Cap Rouge : éditions Presses Inter Universitaires, 1998), 199.

3   Bernard M. Bass and Bruce Avolio, *Transformational Leadership: A Response to Critiques,* in *Leadership Theory and Research: Prespectives and Directions* (Thousand Oaks: Sage Publications, 1994).

4   Douglas Bland, *The Administration of Defence Policy in Canada 1947-1985* (Kingston: Ronald P. Frye & Company, 1987).

5   Canada. *Report of the Somalia Commission of Inquiry* (Ottawa: DND, 1997). Henceforth *Somalia Commission Report.*

6   Ibid.

7   Bass and Avolio, *Transformational Leadership,* 69.

8   Somalia Commission Report.

9   DND, *Military and Civilian Employee Feedback Survey,* Phillips Group/Wyatt Company, Ottawa, 1995, 4.

10   Ibid., 13.

11   Ibid., 15.

12   Lieutenant-General W.E. Ulmer Jr., *Introduction* in Lloyd J. Matthews and Dale E. Brown (ed), *The Challenge of Military Leadership,* xviii.

13   Jim Taggart, *Are You a Transformational leader?* Report by Human Resources Development Canada, New Brunswick Region, March 1999, No. 32.

14   The La Miel school was built in 1997 by members of 430 Tac Hel Sqn during OP CONSTABLE.

15   Taïeb Hafsi and Christiane Demers, *Comprendre et mesurer la capacité de changement des organisations* (Montreal: Les éditions Transcontinental, 1997), 153.

16   Ibid.

17   Ibid., 23.

18   De Pree Max, *Jazz Leadership* (Montreal: Les éditions Transcontinental, 1993), 203.

19   Desaunay Guy, *Comment gérer intelligemment ses subordonnés* (Paris; Bordas, 1984), 8 and 9.

20   Ibid.,1.

21   Bernard M. Bass and Bruce J. Avolio, *Improving Organizational Effectiveness Through Transformational Leadership* (Thousand Oaks: Sage, 1994), 1.

22   *Somalia Commission Report.*

23   Ibid., 3

24   John C. Maxwell, *Développez votre leadership, Un monde différent* (St-Hubert, 1996), 54.

25   Bass and Avolio, *Improving Organizational Effectiveness*, 54.

26   Taggart, *Are You a Transformational Leader?*

27   Ibid., 135.

28   Maxwell, 1.

29   Chantal Leclerc, *Quatre visions du pouvoir dans les groupes restreints, in Leadership et pouvoir, équipes et groupes,* "Gestion des paradoxes dans les organisations" (Cap Rouge : éditions Presses Inter Universitaires, 1998), 76.

30   Ibid., 77.

31   Maxwell, 23.

CHAPTER 8

OBSERVATIONS ON UN STRATEGIC AND INSTITUTIONAL LEADERSHIP – SOUTH SUDAN 2005-2006

Lieutenant-Colonel Mike Goodspeed

At the time of writing this chapter, the United Nation's Mission in Sudan (UNMIS) appears to be succeeding, and in the foreseeable short term, it seems unlikely that the country will relapse into civil war as a result of security related issues. In this respect, the UN efforts have been extremely worthwhile. Nonetheless, working as a member of the United Nations in the Sudan was often a frustrating experience. Despite the organization's maddening sluggishness and its unending flood of administrative gaffes, it should be remembered that it is the only organization that is implementing the peace agreement and carrying out a program to provide a measure of security in that anguished part of the world. By any reckoning this constitutes success. In this respect, my assessments of the leadership issues that I encountered while on duty with the UN should be viewed in light of my support for the notion that the UN is an indispensable international institution doing an essential job. But it is an institution that desperately needs to be reformed – not abolished or marginalized.

It is easy to be cynical about the UN. It is much harder to furnish constructive analysis on the institution's collective leadership. It is in this latter spirit that this chapter has been prepared.

As in any such complex mission, most of the truly significant leadership issues proved to be ones that had elements reaching across the leadership spectrum. Accordingly, in Sudan and in UNMIS one can find relevant topics affecting direct leadership, institutional leadership and the strategic levels of leadership. However, because my position allowed me an excellent vantage point from which to observe much of the internal workings of the military component of UNMIS, I have chosen to restrict the comments in this paper to the strategic and institutional levels of the mission's leadership.

To understand the context of leadership issues in UNMIS, it is important to appreciate the situation that existed there. Violent tensions have existed between northerners and southerners in the Sudan since the days when the Pharaoh's soldiers forayed south from Egypt and into the Sudan to seize slaves for building the pyramids. In January 2005, the Comprehensive Peace Agreement (CPA) ended a civil war that had blazed almost continuously since the country's independence in 1956. Essentially, since the end of British rule, the African south has resisted the Arabic north's attempts to impose on it by force the three cornerstones of Khartoum's Southern Policy: conversion of the south to Islam, imposition of Sharia law and the forcible spread of its Arabic culture. In addition to these longstanding causes of the war, at least, since the late seventies, the African South has demanded its fair share of the wealth generated from the country's newly discovered oil fields. Most of these oil fields lie in the south, while virtually all oil revenues end up with to the government in Khartoum.

The war was a protracted guerrilla campaign that saw the makeshift south Sudanese army pitted against the North's better-trained and equipped standing army. The war, because of its racial and religious overtones, was particularly vicious. It was eloquently, but chillingly, described to me by one northern veteran, a senior Sudanese Armed Forces (SAF) officer as being an endless succession of "just patrolling, ambushes and burning villages; everyday, just patrolling, ambushes and burning villages."[1] This epigrammatic description of the war probably explains quite accurately why there were four million refugees and internally displaced persons (IDPs) – as well as why the actual conduct of the war went largely unrecorded. With the exception of a smattering of armoured vehicles, a few helicopter gun ships and Antonov transports, the war was, for the most part, a very low technology struggle fought almost entirely in Southern Sudan. The war uprooted over four million people and killed over two million all told.[2]

Throughout the years, the north garrisoned key cities and towns along the Nile and clung to a few principal provincial centres that possessed major airfields. From out of these bases, Northern troops periodically patrolled, fighting a kind of "on-again, off-again" search and destroy campaign. For most of the war, the Sudanese People's Liberation Army (SPLA) lived in the bush, in small towns and villages and conducted endless small-scale raids and ambushes. The SPLA's efforts steadily drained the North of its will to fight, and bled it both militarily and economically.

The peace agreement consisted of five elements: the underlying principles for the peace accord; a formula for power sharing in a new, united Sudan, a formula for wealth sharing, particularly the sharing of oil revenues; the resolution of long standing north-south boundary disputes; and, most importantly for the military contingent of UNMIS, which would ensure the implementation of the security arrangements that had to be put in place to provide the country a peaceable future. It is important to note that the peace agreement was not merely a ceasefire implementation process. It was a complex accord that had numerous simultaneous actions, all of which, depending upon how they were implemented, could cement the peace process or lead to renewed hostilities.

By the time the UN arrived in the region, and throughout the duration of my tour, there was still considerable banditry in south Sudan, but organized fighting between the two sides had stopped. The UN therefore was tasked to conduct a monitoring and verification peace support operation and to provide support to assist in acceptance of the Peace Agreement.[3] Force was only authorized to protect UN personnel or to protect civilians and IDPs in circumstances "where they are in immediate danger and if no other protection is available."[4] Despite the elaborately worded references to Chapter VII of the UN Charter, which normally is the authorization for a more aggressive use of force for peace enforcement; the mission was for all intents and purposes a Chapter VI monitoring and verification operation.

My role in the mission was to serve as the Deputy Chief of Staff of the Joint Monitoring and Control Organization (JMCO). The JMCO was a small multinational staff deliberately located well forward in the long-besieged and impoverished city of Juba in south Sudan. The JMCO provided staff assistance to Sudan's most senior military negotiators, and conducted daily preparatory discussions with the Sudanese Armed Forces and the Sudanese People's Liberation Army.

From the perspective of strategic leadership, UNMIS clearly illustrated many of the strengths and weaknesses of the UN. Perhaps the most obvious strategic problem that plagued the mission to south Sudan was that it was a peacekeeping mission, rather than a peace-making mission and it was deployed much later than it should have been. By the time the UN and the international community stepped in, the war had burnt itself out. There is no question that the UN's presence in Sudan has done much to save lives and to ensure that the fighting did not break out again. But as one sage old sergeant said to me the day I got my orders for the Sudan, "That's great sir, but you're leaving three or four years too late."

Of course, the international community cares about such places as the Sudan, the problem is it just does not care enough to actually do anything to prevent or stop these kinds of outrages from taking place. And that, in the simplest terms, is the overriding strategic failing of the UN and its political leaders. It is a failing that colours and permeates the operations and workings of the entire organization. At the heart of many UN operations, the broader international commitment to peace support operations is extremely limited, and there is no willingness to assume any risk on behalf of a cause that is not perceived as an immediate threat to all members of the Security Council.

To put a respectable face on this bald strategic and political failing, there is, in many circles of the UN, an aura of self-important bluster and smug piety with respect to its operations. Fortunately, these attitudes are kept in check by many of the institution's soldiers and civilians who tacitly appreciate the true nature of their deployment and the particular circumstances surrounding the creation of UNMIS. When the highest level of political commitment to an operation is patently hollow, the odds of it being overwhelmingly successful are small. In this case, the UN, for want of member support, had to opt for the much simpler and much more easily achievable task of overseeing a peace agreement long after the real damage had been inflicted. The work the UN was ultimately called upon to do, to oversee and to nurture the peace process, was a useful, necessary and often difficult task, but it was a job that was tragically undertaken much too late.

Notwithstanding the belated strategic context of the mission, the quality of the mission's strategic leaders was high. For all intents and purposes, there were, in reality, only two truly strategic leaders on the ground: the Special Representative to the Secretary-General (SRSG), and his Force Commander.[5]

The SRSG, Mr. Jan Pronk of the Netherlands, was perpetually criticized by the Sudanese press, and, dismayingly by far too many of his civilian UN employees for being insufficiently diplomatic – a trait most soldiers in the mission found to be a refreshing and desperately-needed attribute in the UN's official but nonetheless hypocritical institutional environment of moral equivalence. Despite such differing opinions of the political strategic leadership, the fact remained that, unless the most senior UN official spoke plainly about linking the crisis in Darfur with the wider prospects for peace in the south, all efforts toward a lasting peace in Sudan would have become a charade. In this respect, the SRSG displayed a high degree of honesty, candour and integrity – all of which are fundamental preconditions for strategic leadership. But unfortunately, they would seem to be traits that have all too frequently been sacrificed at the most senior diplomatic levels in favour of short-term expediency.

The Force Commanders (there were two of them during my tour, Major-General Akbar from Bangladesh, and Lieutenant-General Lidder from India) were both highly principled, shrewd, tough, energetic, experienced and capable soldiers. Given the mandate they were tasked to implement, they both wrestled, for the most part successfully, with problems that were not of their making; and given the constraints imposed upon them, they managed to keep the peace process from floundering. Both officers moved the peace process forward. In doing so, both Force Commanders displayed an infectiously optimistic outlook which they translated into an endlessly patient but practical vision for implementing the CPA. And both men, in their own very different ways, were successful in motivating a highly diverse military organization.

It is worthwhile noting that during the earlier days of UNMIS, the UN had certainly done a good job in choosing its strategic leaders for the Sudan. And while the collective UN leadership in New York is readily singled out and censured for its shortcomings, it must be given full credit for wise decisions in its leadership selection. Leadership selection is a critical skill and one that can compensate down the line for other institutional shortcomings. The reverse, of course, is that when poor strategic leaders are placed in positions of power, other problems are invariably amplified.

Strategic leadership in a peacekeeping mission differs from that of a traditional war fighting campaign. In the context of a Chapter VI peacekeeping mission, the key features of the commander's plan are, for the most part, developed for him in detail by the terms of the peace agreement. In implementing a sequence of actions that have been carefully hammered out by those who preceded him at the peace conference, there is little scope left for broad changes to the plan. Nonetheless, the Force Commander can do much to influence the pace and tone of operations. He can develop personal relationships with key players on both sides, and through these relationships he can cultivate and influence the long-term expectations of all parties. Of equal importance, the Force Commander sets the tone as to how aggressively and doggedly the military component of a peacekeeping mission reacts to threats to the ceasefire. In this respect, his strategic influence is vital to the success of the operation. Often, the personal influence of the Force Commander can have an immediate calming influence when there are breaches of the ceasefire; or, as is often the case, when over time tension builds between the two sides. On a long-term basis, his personality and influence does much to shape and develop the attitudes of the belligerent parties. And in this respect, trust, goodwill and a cautious respect for the peacekeeping force must be regarded as strategically valuable assets.

Strategic commanders in a multinational peacekeeping setting are faced with some unique difficulties in terms of welding together a team that will be responsive to his objectives and operating style. Force Commanders are usually rotated in and out of their positions on a yearly or bi-yearly basis; and this, in a multinational and multi-cultural setting, in turn creates difficulties of continuity and focus. Also in the UN context, a strategic leader will likely have little control over who is on his staff, what position they will hold and what training and preparation they have undergone. This sharing of the peacekeeper's strategic responsibilities with the UN staff in New York reduces the commander's flexibility and places demands on his creativity and time management skills.

Fortunately, the staff of the Directorate of Peacekeeping Operations (DPKO) has, for the most part, managed to ensure that most of the key positions in their field headquarters are filled by reasonably competent and experienced individuals. Nonetheless, the issue of quality in key staff officers remains for obvious reasons, an extremely sensitive matter. For if the international selection pool is small and is based solely upon current availability rather than well-established national commitments, the likelihood of having incompetents in a key staff position rises dramatically. And this, in turn, affects the character and prospects of success for the entire mission.

Thus, one of the long term reforms that Canada should champion in DPKO is the institution of practical and clearly defined quality control standards for officers who are assigned to important staff positions for UN operations. This, of course, presupposes that Canada is not only willing, but is also capable of providing significant numbers of our own most promising and energetic officers for staff service in support of UN operations. Given our constraints on the number of formed units we can supply to such endeavours, Canada, with only a modest increase to its existing investment, could do much to enhance the UN's institutional effectiveness, strengthen the key staffs of the UN's most senior commanders, and, at the same, time re-build our profile in an important international institution.[6]

The strategic demands of the most senior leadership in a UN peacekeeping operation vary in degree rather than kind from those of a coalition commander in a war-fighting operation. The nature of command, the degree of authority, the diplomatic nuances, the need to project a strong personal identity, the degree of trust as well as the expectations that the commander has of his staffs, formations and units are all in one form or another different. Some might argue that different kinds of personalities and backgrounds are required for a strategic leader in peacekeeping. However, what is common to the war-fighting strategic leader and the peacekeeper is that both leaders require a strong military background and depth of experience in order to make the kinds of decisions that will be required of them. And, not of least importance, a peacekeeping general officer, like the most junior military observer, must, at all times, have military credibility with the warring parties. The issue of credibility is unfortunately often overlooked in discussions regarding the nature of Chapter VI peacekeeping.

One of the overarching and distinguishing features of UN operations is that they are all too often "ad hoc" affairs. This is, in part, driven by the fact that each and every UN mission is dependent upon the goodwill of member states to volunteer troops, equipment and funding. As was the case in the Sudan, pledges that are made are often not kept. Deadlines are frequently ignored, and quality and quantity assurances are routinely disregarded. This built-in uncertainty, in addition to other intrinsic disadvantages, drives an environment of caution and circumspection that affects every level of leadership. Combined with the fact that the UN is, at heart, a multi-cultural organization infused with, what to all intents and purposes, is a voluntary outlook, many individuals within the institution display a work ethic comparable to that found in a charitable fundraising association. There are thousands of enormously hard working people in the UN, but there is also a disturbingly high percentage of dead weight. In these circumstances, the strategic leader's range of action is constrained, and the issue of command presence takes on a unique connotation. The Force Commander is looked upon to set the organizational tone, and this is not an easy thing to do within a cumbersome multi-cultural force that displays a wide range of attitudes on matters such as personal work ethics, loyalty, commitment, and cooperation. Moving an agenda forward and working around the expectations and attitudes of such a diverse force is not an easy task, but it is an inescapable element of the strategic leader's environment in the UN.

Perhaps some of the most compelling and insightful examples of strategic leadership that we should study are to be found amongst the warring parties themselves. Unfortunately, detailed knowledge of the parties to the conflict in the Sudan proved to be one of the areas in which we were inadequately prepared. Sudan was an excellent setting for examining leadership in the context of sub Saharan Africa and the Arab world. One of the most obvious examples of strategic leadership is to be embodied in John Garang.

John Garang, the charismatic leader of the SPLA died in an airplane crash shortly before I arrived in the country, but even in death, his influence upon the traditions and behaviour of the south Sudanese military and political leadership was profound. Garang was, in his lifetime, and in the period following his death, accorded legendary status. Unfortunately, other than what can be gleaned from overly simplistic popular press clippings, we know very little about the man, what made him so admired, what were the key elements of success in his military and political leadership style and what circumstances enabled him to fire the imagination of so many people. Likewise, with the leadership of the Sudanese military, despite our proximity in working with them, we (and I would include virtually everyone I met in the UN in the category of "we") still know very little about their leaders and their general leadership milieu.

There is likely to be turmoil in Africa for some time to come. Accordingly, the Canadian Forces (CF), and the rest of the world, should learn considerably more about the belligerent leaders and their situations than we do at present. Perhaps deliberately focused, well-considered post-graduate study at the Royal Military College of Canada and the defence forum universities will provide us a more useful insight into these men and their conditions. Such work would also provide us a much better understanding of the nature of modern Africa. It would give us a greater appreciation of modern guerrilla movements; and, it will increase our awareness of Islamic armies. And, perhaps of more immediate benefit, we might also gain some insight into some of the tangential issues surrounding the Greater War on Terrorism.

An examination of "Institutional Leadership," within UNMIS probably reveals more about the way UN operations are conducted than does any other area of study. The CF defines institutional leadership as that function "concerned with developing and maintaining an institution's strategic and professional capabilities," and "which creates the conditions for operational success."[7] In its broadest form, institutional leadership is about the kind of culture that an organization perpetuates and how that culture contributes to the success or failure of the institution. Because leadership in this "institutional" sense is largely defined by the organization's culture, it is an amorphous concept wherein the temperament and personality of the organization is shaped by the collective character of its members.

As is recognized in our own leadership doctrine, institutional leadership issues invariably bleed into strategic issues, and the dividing line between the two categories is rarely distinct. In the Sudan, there were two major kinds of institutional issues that presented themselves. The first can be classified as those issues that were pre-determined by the nature and personality of the UN itself. The second are those important policies, plans and orders that originated almost quasi-anonymously as a result of numerous inter-departmental iterations, reviews and approvals within UNMIS.

In a UN peacekeeping mission, which is by nature a highly rule bound organization, within which there is frequent personnel turnover, leadership, at the institutional level, tends to define the character of the mission. The character of UN Chapter VI missions are for the most part defined and constrained by a web of negotiated decrees, and administrative rulings as well as by the logistic, security, mobility and tactical limitations to action that are imposed upon a force deployed for monitoring and verification tasks. Only periodically and with enormous effort do strategic leaders manage to leave a distinct and lasting personal stamp upon the conduct of operations.

Unlike the conduct of war, where lethal combat demands that commanders imaginatively and strenuously strive to destroy their opponent, in Chapter VI Peacekeeping Operations, the Force Commander's latitude for innovation and creativity is more rigidly circumscribed by rules and regulations. The mission's objectives, as well as almost all its associated courses of action have been clearly determined in the negotiation stage of the peace process. In this respect, there is little original planning that goes into designing the operational concept of a Chapter VI mission.

This was most evident in the collection of documents that spelled out how UNMIS was to operate. In Sudan, the Force Standing Orders, most of which were photocopied from other missions, ran to dozens of volumes, all of which were maintained on the shared drive. This was an appropriate place for the material, given the austere conditions in which most of the deployed units initially found themselves, and where few sub-units would be able to manage the scores of detailed instructions that made up the Force's Standard Operating Procedures and Standing Orders. At first glance this may seem to be a harsh assessment of the officious nature of the UN. But, in fairness, modern militaries are, by comparison, quite literally supported with libraries of legal, technical, training, and administrative instructions. However, in this respect, the difference between the UN and most modern armies is that the United Nations must assume that within a hastily assembled multicultural organization, the baseline of shared understanding of how a force operates is at a lower level than one would find in a more permanently organized and stably funded standing military force. In tacit appreciation of the fact that, in such an organization, one that has so many different cultures and where nothing can be taken for granted, things that would normally be left unsaid and assumed need to be spelled out. Hence, there is a requirement for precise instruction. And in the context of the UN, this translates into volumes of detailed directives, many with which personnel in the field lack familiarity or even awareness – and this situation was a constant source of friction between the various headquarters staffs and units deployed in the sectors.

One key aspect of the institutional leadership within UNMIS that was most evident in south Sudan was that UNMIS was created as an "integrated peacekeeping mission." From an institutional perspective, this meant that working alongside the armed forces there were large numbers of civilian police, UN Agencies of various descriptions, scores of non-governmental organizations (NGOs) and the UN Administrative apparatus.

For many, this "new" approach to peacekeeping did not appear substantially different from what had been happening for years in far flung places such as the Balkans, East Timor and the Congo. In the Sudan, especially in the headquarters in Khartoum, there was considerably more evidence of integration than had heretofore been the case, and the military presence in Khartoum was notably smaller than the civilian element. Within the various sectors in the field, this ratio was reversed.

In a situation such as in Sudan where the shooting has stopped, an integrated force is logical for two reasons. The first is that the peace process had to be cultivated on several planes, not just on a military dimension, and this requires the close integration of a very wide range of organizations. The second reason was much shorter term and more practical. Many of the more intrepid NGOs and United Nations Agencies (one of the largest and most notable being the World Food Program (WFP) had remained in south Sudan throughout the most dangerous periods of the war. These organizations had a tremendous amount to offer in terms of local knowledge, intelligence and an understanding of how the two sides operated and thought.

Notwithstanding this, it appeared that there were very few of these individuals actually recruited into UNMIS when the mission was raised. Because UNMIS was created after the cessation of hostilities, it was developed "from scratch" and recruiting to fill key positions came largely from within the UN Secretariat's career stream as well as from requests to donor nations from the Department of Peacekeeping Operations in New York. The result was that, without having a leavening of individuals who had experience in the region put into key positions, the learning curve for those tasked with setting up the mission was considerably steeper than might otherwise have been the case. This meant that those UNMOs, national military contingents, and aid organizations that deployed into the field in their respective sectors did not have access to the kinds of useful local knowledge that would have helped them to become operational as quickly as possible.

As an institutional leadership problem, the results of this kind of "grass roots" coordination had numerous practical consequences for the mission itself. The initial deployment of UNMIS had been much slower than planned, and, despite having the availability of a highly professional and competently led planning staff from NATO's SHIRBRIG Brigade, the mission suffered from a number of avoidable administrative problems.

The UNMIS deployment was hamstrung by the failure of some of the key troop contributing nations to deploy as promptly as they had initially promised. Moreover, for a variety of reasons, essential components of the Force, such as medical units and tactical aviation units, took much longer to be brought into theatre than had originally been envisaged. Such problems do not exist in isolation. Logistic problems were compounded by other issues; for within both the military and civilian elements of UNMIS there were few people who actually had experience operating in this particular part of the world.

This combination created avoidable and predictable problems for the mission. For example, some of the staff deployment estimates miscalculated the nature of the area's rainy season. The length and intensity of the rains in conjunction with south Sudan's clay based soil and the mine problem, meant that roads were frequently impassable. With inadequate helicopter lift, logistic efforts, no matter how well intentioned, could not proceed as envisaged.

In this respect, institutional leadership failings were a culmination of the collective errors and miscalculations of a host of nameless staff officers, bureaucrats and contractors from several donor countries. This issue, combined with a less-than-adequate understanding of the region, meant that, despite having access to a first class planning staff, the Force's deployment was slow and inconsistent.

As a result, the delayed UN build-up helped foster a sense of disappointment and a loss of confidence in the UN's capabilities amongst the Sudanese people. This disillusionment was also seriously compounded by wildly unrealistic expectations that many south Sudanese harboured with respect to the UN. Many of the local civil and military leaders assumed that the UN was going to build roads, hospitals and schools as well as "kick-start" their economy. Thus, systemic logistic problems and a failure to understand the information needs of the population, influenced the tone of military negotiations right down to the local level. Handling these information and perception problems, in the context of a "joint mission," was the responsibility of the UNMIS Public Affairs Office, which was inadequately staffed and under-funded and experienced its own set of frustrations in coordinating the work of the mission's military and civilian agencies.

The foregoing is but one example of some of the interconnected and systemic coordination problems that were inherent in UNMIS joint operations. These issues demonstrate a kind of leadership problem that was often not glamorous or readily apparent. Nevertheless, the ability to effectively overcome these problems and many more was essential to helping to establish confidence in the peace process and to building a basis for the re-establishment of civil society.

In retrospect, the handling of these highly inter-related issues became institutional leadership problems, problems that required the concerted efforts of numerous different groups. In the case of the UN, it would be pointless to try to identify one organization or group of individuals who had the ability to galvanize the UN into becoming a responsive and energetic institution. In the future, fixing shortcomings similar to these is only likely to take place if the character of the larger institution changes. And this is a long and difficult process.

Although within UNMIS there was a shared sense of purpose, and personal relations between UN troops, agencies and departments were on the whole quite good, the operational integration of these organizations was not well developed, nor was there a sense of coherence in joint undertakings. There were unquestionably dynamic, hands-on leaders across the organization, but their efforts were all too often blunted and frustrated by reluctant and uncertain bureaucrats and staff officers who were interpreting policy manuals and administrative instructions. And, perhaps the single most important lesson is that diversity is a wonderful concept in terms of enabling people from all backgrounds to compete for positions of influence, but key decision makers should be selected and employed on the basis of their abilities. Individuals who stalled and quibbled and who refused to make decisions from a sense of entitlement, fear of failure, or arrogance were responsible for much of the lethargy in the UN. However, if the UN could hire and retain people based upon their demonstrated abilities, the institution would be much stronger. If hiring practices were based upon ability, it is also unlikely that the UN's middle ranks would become exclusively populated by individuals from the First World as is feared by so many.

In my experience in the Sudan, there were numerous capable and dedicated leaders who came from every corner of the Third World. Equally, there were incompetents who came from Europe and North America. The first step the UN needs to take in addressing its institutional leadership problems is to have the courage to admit the nature of the problem.

Although Chapter VI peacekeeping has been a reality for some time, one should remember that this form of joint undertaking is still in its infancy. And viewed in this light, it is worth bearing in mind that over time, the UN will improve. Perhaps one model the UN could use to help hasten improvements in their conduct of joint operations is that provided by the NATO-led coalition in Afghanistan. Or conceivably, the UN could adopt the example of some of the world's most effective large-scale multinational corporations. There are several effective models from which to choose.

Despite the scale of the problem facing the UN, the situation is not entirely bleak. There are encouraging signs for improving the organization's institutional culture and for developing the quality of its middle level leaders. For example, the strategic leaders the United Nations chooses for its missions are usually highly competent people, because they are selected from amongst the planet's best and brightest men and women. This has been a successful practice, as both leadership selection and retention have been based primarily upon competence, and only secondarily upon national origin and other factors. If the UN is serious about injecting vitality and competence into the institution, it should begin by following the same practices in selecting its middle leaders that it does for its strategic leaders.

ENDNOTES

1   Author's interview with a Sudanese liaison officer attached to the UN Joint Monitoring and Control Office, Juba, Sudan, 12 December 2005.

2   The casualty rates for the war are conceded to be rough estimates. The figures of two million dead and four million displaced persons are figures that have been generally accepted by the international press as well as the UN, although the nature of the country, its record keeping and its penchant for secrecy means a precise accounting of these figures will likely always be impossible.

3   "UNMIS Military Roles Tasks and Responsibilities," UNMIS Chief Operations Officer's Briefing 20 October 2005.

4   Ibid.

5   The Deputy SRSGs (Humanitarian and Political Affairs), the Chief Administrative Officer and the Police Commissioner, although of equivalent rank and status to the FC and despite being vitally important to the mission's success, were more closely involved in implementing the SRSG's vision and their efforts are more rightfully categorized as being institutional in nature. In terms of far reaching, over the horizon decision-making I have categorized the Force Commander and SRSG as being strategic leaders.

6   Canada has been quick to supply military observers and we have undoubtedly consistently fielded high quality individuals. However, many of these individuals are usually selected on an as available basis and selection for service as an UNMO is often the result of fortuitous circumstances as well as astute and understanding supervisors rather than through deliberate planning.

7   Canada, *Leadership In the Canadian Forces: Conceptual Foundations* (Kingston: Canadian Defence Academy, 2005), 131.

CHAPTER 9

CIVIL-MILITARY OPERATIONS IN COMBINED FORCE COMMAND — AFGHANISTAN

Colonel François Vertefeuille

As the Combined Joint 9 (CJ 9) staff officer (civil military operations) of Combined Force Command Afghanistan (CFC-A), I was responsible for two different files: developing strategic policy for the U.S. Provincial Reconstruction Teams (PRTs) and supervising CFC-A headquarters (HQ) liaison officers (LOs). There were about a dozen of these officers at the beginning of my mission, and the number increased to about 20 after the rotation of American personnel in June 2005. My other mission, although it did not appear in my official job description, was to act as the senior Canadian officer at HQ. As such, I ensured the well-being of three other Canadians, all majors (who worked in different sections than I did) and acted as the main contact point when the Deputy Chief of the Defence Staff (DCDS) in Ottawa wanted to obtain information concerning CFC-A activities. As a staff member who was part of the Commander's Conference, I was in a good position to keep my ears open for any information that I judged to be of interest to National Defence Headquarters (NDHQ), which was planning Canada's future activities in Afghanistan. My direct access to leaders at HQ (Commander, Deputy Commander and Chief of Staff (COS)) facilitated the transmission of information between CFC-A and the Canadian Forces (CF).

Being a signals officer, I was completely taken by surprise when I was assigned to this position. This occurred due to various reasons. As the deployment date for the appointment was approaching rapidly, the Chief of the Land Staff (CLS) was still seeking a candidate to fill the position. My name was submitted and accepted. It took me three weeks to wrap up some critical files in my position and to take a little pre-deployment training, one week's leave, and one last week to finalize some personal administrative matters, then I was off to Kabul. Five weeks elapsed between the time I was named to the position and my departure date. That meant that I had no civilian-military (CIMIC) training to familiarize me with the duties that awaited me. The mission was going to be challenging.

BUILDING A TEAM

Upon my arrival in-theatre, I joined a very diverse, but well-established, team already in place. One of the reasons for the exceptional diversity of this team (I will return to this point later) was that the majority of operations planning and control activities were carried out using the American SIPR, a "SECRET—US EYES ONLY"- level system. It was not until around the midpoint of the mission that the British and the Australians received authorization to access the system, and at the very end of my stay the Canadians were granted the same privilege. CFC-A is a coalition comprising personnel from a number of nations, and the restriction on access to the main information system was a major constraint. Because civil-military operations do not require access to a classified system, many military personnel from countries other than the United States (U.S.) were assigned to that section. On my team were six Romanian, one Albanian, one Polish, one British and a dozen American soldiers. Not all countries necessarily share the same vision of how to operate in Afghanistan, so familiarizing myself with the American vision, making it understood, and above all getting the non-American officers in CJ-9 to accept it was sometimes a considerable challenge. I truly believe that, when we broached sensitive subjects, the fact that it was a Canadian defending the HQ position probably made it easier to solve the problem of ensuring that such messages were accepted. I had to personally familiarize myself in depth with American policies and directives. Let me be very clear about this: even though the HQ was that of a Coalition (Combined), the U.S. provided well over 90 percent of HQ personnel, troops under lower HQs, and funds needed to ensure the success of operations. We were thus operating under American directives. In addition, the higher HQ for the Coalition was Central Command (CENTCOM), a U.S. command.

When a group of armed forces members from different nations forms a minority within an organization such as CFC-A HQ, unusual situations can arise from time to time. For example, for information security reasons, the CJ-6 regularly checked the software installed on the computers. One of these checks uncovered a chat program used by the Romanians for communicating with their families. Because the software was unauthorized, it was removed from the computers (which were connected to HQ's unclassified system (NIPR)). The Romanians considered this initiative an affront. When the senior Romanian officer requested a meeting with me and let me know that his countrymen felt harassed and discriminated against, I understood the importance of the situation. I discovered that the Americans could use the telephone system at HQ to communicate with their families back home but the Romanians could not do the same, and that the software was their only means of communication with their loved ones in Romania. The problem was a perceived injustice.

Unfortunately, a perception is often more difficult to correct than a real problem. That was one of the occasions on which my knowledge as a signals officer stood me in good stead. I was able to explain to them that in Canada, similar controls on software were also in place, and that I myself had ordered such action in the past. As a Canadian officer, my credibility and impartiality were not questioned. When I discussed the problem with the CJ-6, he and his staff understood the importance of finding an alternative solution that would allow the Romanians to stay in touch with their families, while complying with information systems security instructions. A technical solution was found that enabled the Romanians to phone home.

I realized during this incident that the systems in place were meeting the needs of the majority of HQ personnel (in this case, the American military community), but that the situation of armed forces members from other countries was not known to or understood by the authorities. Such situations can give rise to highly charged incidents, as the minority may quickly jump to conclusions and feel slighted. I believe that my 25 years of experience as a commissioned officer in the CF helped me get the facts and recognize that the problem was mainly one of communication between a majority and a minority group. The challenge was not really to solve the problem, but rather to help both sides understand the other's needs and see that their perception of the situation was inaccurate. The problem was solved through technical knowledge. The perception issue was solved using leadership skills.

When I arrived, the American officers were in the eighth month of a one-year operational tour. They had established a routine, and their objectives had been clearly identified. After a few weeks in-theatre, I developed a good understanding of the liaison officers' (LOs') activities. A few discussions with the COS made me realize that these activities had not necessarily produced the results that leaders at HQ were expecting. Indeed, LOs had been told in the past that they were not spending enough of the money at their disposal to help the different ministries of the Afghan government. They had quickly made some changes and, under the directives of the senior LO, generated an impressive number of projects. The CJ-9 annual budget of U.S. $23 million is sufficient to provide significant assistance to numerous ministries, which all too often had to be completely reconstructed after 30 years of war.

I came to the conclusion that my LOs were serving the Afghan government much more than they were CFC-A HQ. I wanted to redirect their efforts toward information-gathering, in order to help HQ gain a better knowledge of the situation in the different ministries. I wanted to transform the CJ-9 into an "info-centric" section rather than one focused on generating projects. That was easier said than done. Although the LOs were very comfortable with their activities, not everyone was happy about the arrival of a new CJ-9 with new priorities. It is important to understand that it was difficult for the LOs to make the employees at the different ministries, particularly the ministers themselves and their numerous deputy ministers, understand that the funds available to them now came with more conditions attached. I wanted the LOs to use these funds to signal CFC-A HQ's good intentions, but the main goal was to exchange information with the ministries in order to help them restructure in a way that would enable them to fulfil their responsibilities. We were progressing slowly in that direction when there was a change of command in CFC-A.

We are all familiar with the principle: new commander, new vision, new directives. When I gave a presentation to the new Commander on CJ-9 and my objective of changing the LOs' main focus, the Commander applauded my initiative—and issued me directives on the use of funds that were even more restrictive. We were generating numerous projects that were social in nature (e.g. repair of schools and hospitals, assistance to women and children) or would help ministries run better (e.g. computer purchases, classrooms, English and computer courses), but the Commander ordered me to accept only security-related projects and assign the LOs only to high-payoff ministries. These new directives created enormous changes in the LOs' way of working, and these changes caused considerable discontent.

I had to deal with a very difficult situation in which the officer in charge of the LOs did not support my new objectives and even went so far as to tell the LOs not to pay too much attention to my directives (which, I must emphasize, came directly from the Commander). The fact that the officer in question held the same rank as mine did not make the situation any easier. Certainly, transformational leadership is the best way to build team spirit, but this situation required authority. Less than 48 hours later, the individual was headed back to the United States.

From that moment on, my LOs were much more cooperative. I understood the expectations of the numerous ministries these officers had supported for almost a year, and the difficult situation they were all in, and I knew there was only a month left before they would be replaced by a new team of LOs from the United States. As a result, I succeeded in convincing the Commander that it would be preferable to quietly withdraw from the ministries not on the "high-payoff" list, while limiting new projects to those related to security, as he had ordered. Projects for which funds were already committed were transferred to a project management team, which followed them through to completion.

In this way, I succeeded in getting my new objectives accepted by the Commander, but I had to defend my personnel and make it clear to him that we could not make a 180 degree turn in the space of a few days without causing HQ to lose the credibility and respect that it had earned through a year of hard work by a dedicated team of LOs. It was quite a challenge. I also had to convince the LOs that the Commander's aim was to advance the ultimate mission of CFC-A: to create a stable environment in which the Afghan government could ensure the safety of the population.

A CHANGE OF TEAMS

In the U.S. Army, it is the Civil Affairs (CA) battalions that have expertise in civil-military operations. These units are part of the Reserve and are mobilized several months before a deployment. The change of CA battalions took place about four and a half months after my arrival. The unit trains in the United States and prepares and organizes itself according to the information obtained from in-theatre reconnaissance. Even before the members' arrival, they are already a team. For this reason, the CA officers assigned to CFC-A HQ thought they had a good idea of what awaited them. They knew that a Canadian colonel was to be their section head, but they were very surprised to see the extent to which the CJ-9 mission had changed as a result of the reorganization ordered by the

Commander. The new CA battalion team was made up of about 15 officers, of who seven were colonels. Together with the senior American officer, I had to accurately identify the particular skills of each LO in order to assign them to the high-payoff ministries in such a way as to take maximum advantage of the knowledge they had acquired in their civilian careers.

Once again, I convinced the Commander to wait until the new LOs arrived before restructuring the section according to his directives. That way, it was easier to explain to the new arrivals the objectives set by the Commander and let them get involved. The previous team, after almost a year in-theatre, would have had much greater difficulty changing its objectives and procedures and adapting to them, after becoming set in its own work habits. As a final task before their return home, I instructed them to withdraw slowly from the ministries that had not been deemed "high-payoff," to complete as many projects as possible and to explain to their ministries that only security-related projects would be considered from then on. The new team arrived and took up its duties without having to undergo a change of direction. I was lucky; the rotation could not have happened at a better time.

Being in charge of the activities of seven colonels was sometimes a delicate task. In addition, they were all trained as CA officers, while I personally had received no such training. Only my four months of experience at HQ gave me a better understanding of the situation, as well as a very clear idea of the Commander's expectations. That experience, together with the new arrivals' professional knowledge, enabled us to quickly build an effective new team. Although I was in charge of the section, I had to pay attention to ideas and suggestions from my personnel. For their part, they were well aware that I was the one who was ultimately responsible to the Commander for the section's performance.

LEADERSHIP WITH OUR NEW ALLIES

Working with personnel from a number of different countries can present a particular challenge. Forces members from the former Eastern Bloc countries are accustomed to a very different leadership style than that which prevails in the Canadian Forces. I realized that the members of the Romanian, Polish and Albanian forces were almost fearful of coming to consult me. I had the impression that for them, asking for advice or more specific information about a directive was almost an admission of defeat and incompetence. Building a functional team was quite a demanding task in the beginning, given that I had to keep an eye on the activities of certain individuals who I knew preferred to venture into the unknown rather than consult me. Merely entering my office required them to display excessive politeness and offer myriad apologies for "disturbing" me. For example, it was very difficult to make the two Romanians, who looked after the coffeemaker, understand that a Canadian colonel was perfectly capable of pouring himself a cup of coffee!

Although I tried from the beginning to explain to them that I was always available to speak with them, that they were not disturbing me, I believe it was through seeing the working relationship I established with the American and British personnel that they came to understand that responsibility and leadership do not necessarily equal dictatorship. Within a few months, the situation had greatly improved. I often wonder whether, after repatriation, they tried to establish similar relationships with their subordinates. If military personnel from these countries serving in Afghanistan are representative of their armed forces as a whole, it will take several years before they are able to apply the "mission command" principle. However, I do not want to leave the impression that our system is "the best" and that theirs is "not functional." I do not consider myself enough of an expert on leadership to make a judgment on that point. But I definitely observed a very different leadership style on their part, and my observations confirm the truth of all the lectures and presentations we received during our training on the former Soviet Bloc countries.

THE BIGGEST CHALLENGE

Working in an almost entirely American organization, having other colonels under my command and managing personnel from former Eastern Bloc countries were all considerable challenges. But the biggest challenge I have ever faced was co-chairing the working group on the Provincial Reconstruction Teams (PRTs). The first Commander of CFC-A set up a PRT Executive Steering Committee. This committee was co-chaired by the Minister of the Interior in the Afghan government and by the commanders of CFC-A and the International Security Assistance Force (ISAF).

All countries that had PRTs in Afghanistan and those with serious plans to deploy one in the near future were invited to be members. The person sent to sit on the committee was normally the country's senior representative, often the Ambassador. The committee's objective was to give strategic direction for PRT operations. To prepare for the committee's meetings, a working group (WG) was formed. Each country represented on the committee was invited to work with the WG. The Chair of the WG was a brigadier-general on the Afghan police force, who reported to the Minister of the Interior on all issues related to the PRTs. But the Chair had two handicaps. He knew very little English, and he confessed to me during a one-on-one meeting that cultural norms prevented him from speaking openly to a large number of people about the problems he was facing.

However, WG members had no difficulty raising problems in the forum. On the contrary: co-chairing—and even, by default, chairing—the WG was one of the most difficult experiences of my career. To give an example of the kinds of difficulties I encountered, the first five WG meetings following my arrival in-theatre were entirely devoted to discussing which topics the WG did not want to discuss. It took me three more weeks to get discussions started to identify a series of topics that were acceptable to the members. The WG was seriously dysfunctional, for a number of reasons.

First, I had the impression that several members of the WG were diplomats at the beginning of their careers who wanted to seize the opportunity to speak for their country, in order to show their superiors that they had succeeded in defending their national interests. Second, some of the group's unofficial leaders were convinced that non-military organizations, such as a member from the United Nations Assistance Mission in Afghanistan (UNAMA) or an embassy representative had the right to chair or co-chair the WG. They were quick to criticize, but when we invited them to take on the responsibility of co-chairing, no one volunteered, although they continued to complain that military personnel were in charge of the WG. And lastly, some nations were very reluctant to discuss their PRTs' operations.

That last point is very important. Within the CFC-A coalition, all the PRTs came from the United States, except the one from New Zealand. The CFC-A Commander's directives were law, and the PRTs had only to carry them out. Within ISAF, the situation was totally different. More than 10 different countries provided a team, and each country had different rules of engagement (ROE). Because American CIMIC doctrine and rules of engagement were different from NATO's, it was difficult to compare the teams' effectiveness.

By definition, the PRTs are temporary entities put in place until the Afghan government can take charge of the tasks they are carrying out. The American Secretary of Defence, Donald Rumsfeld, told my Commander clearly that he wanted a transition plan developed to determine when the American PRTs could be withdrawn from Afghanistan. The Commander in turn quickly issued directives on this. In order to decide when a PRT could transfer its functions to the Afghan provincial authorities, evaluation criteria had to be developed. At that point, we reached an impasse. As mentioned earlier, CFC-A could quickly modify PRT operations. Since ISAF had PRTs in the two most stable regions of the country (the north and the west), and the Americans were operating in the regions where there was a great deal of enemy contact, the Commander could easily order the PRTs to provide direct support to forces involved in combat operations, even though doctrine strongly recommends keeping these two types of activities separate. The ISAF Commander could direct PRT operations but had to stay within the limitation of each respective participating country's national ROE. Any deviation had to be approved by the national capitals, and getting this approval was a long and complicated process. Therefore, the WG members representing nations whose ROE were much more restrictive than the Americans did not want to develop evaluation criteria for the PRTs, for fear that their teams' operations would be judged less effective than those of other nations.

Thus, I found myself between a Commander who, under pressure from the American Secretary of Defence, was expecting quick results, and a WG that did not even want to hear about the subject. We approached the problem from two angles. First, we began developing criteria "off-line" with those who were interested in participating and those who had an interest in ensuring that their opinions were taken into consideration when developing such criteria.

The PRT Director for ISAF and his staff were excellent allies on this issue. They realized that even though some nations were afraid of these evaluation criteria, they remained indispensable in the long term. In addition, we succeeded in modifying the criteria so that they evaluated the prevailing conditions in a province, and not the effectiveness of the reconstruction teams. This made them much more acceptable to many nations. Lastly, we managed to convince UNAMA to oversee the work of developing these criteria, which gave them more of a civil than a military flavour, one that was more objective and based on internationally recognized studies and projects.

Many other events of this kind occurred during my stay and while I was co-chairing the working group. Transforming the Commander's very clear directives into work proposals for members of an international diplomatic community and meeting the deadlines imposed presented quite a challenge.

Amidst all this manoeuvring, it must be remembered that the ultimate goal of the PRTs is to help the Afghan government extend its influence beyond the capital. The real problems still needed to be aired and corrected. The Afghan Director of PRT Operations, as mentioned previously, had encountered cultural problems that prevented him from broaching certain subjects requiring the WG's attention. I tackled this problem by paying an unofficial weekly visit to the general. While exchanging pleasantries over Afghan tea, I succeeded in establishing a friendship with him as an individual and was able to put him sufficiently at ease that he would tell me about his problems. Thus, upon my return to HQ, I could take action to correct certain situations, or inform the ISAF PRT Director if they concerned one of his teams. Later on, the latter joined me during my weekly visits. Once a problem was identified, if it concerned the PRTs in general, the ISAF PRT Director and I would come up with a plan for raising the matter with the WG members in a way that would make it an acceptable discussion topic in their eyes, or we would address it ourselves immediately. The vast majority of real problems were settled outside the WG, between the Afghan general, the ISAF PRT Director and myself.

Serving as liaison between a military coalition HQ and a group of individuals from diplomatic and international aid organizations was one of the most difficult tasks of my entire career. Persuading this group of individuals, some of whom were overtly hostile to the military, to discuss matters related to operations of the PRTs (which are military units with a humanitarian objective) was extremely laborious.

FATIGUE

Working between 100 and 110 hours per week for four months straight eventually drains one's energy. It was necessary to find a way to recharge, not only physically, but also mentally. I kept myself in good physical shape by jogging regularly two to five times a week, depending on my schedule. I quickly realized that those who were not in good physical condition had a great deal more difficulty concentrating after a long day's work and frequently caught the flu. Since Camp Eggers (the CFC-A HQ camp) did not offer an ideal environment for jogging, I tried to work out on a stationary bike as regularly as possible. That helped me stay in acceptable physical condition. Some individuals were able to remain continuously focused on work without ever feeling the need to take a breather. I humbly admit that that was not true for me; I needed a change of pace from time to time. Working out was a big help, but I also set aside half an hour just before bedtime to read and listen to music. Before shipping out, I had prepared a dozen MP3 CDs, and I brought them with me, along with a good personal stereo. That was how I managed (hopefully) to maintain the mental and physical balance required to keep up the pace. Technology was also a big help, particularly the Internet. I had Internet access at my workstation, which allowed me almost direct contact with my loved ones. Because I was based at the American camp, and given the time difference and the restrictions on travel after dark, it was impossible to be at Camp Julien to make phone calls at times when I could reach people in Canada. But thanks to the Internet, I was able to stay in close touch with my loved ones and be reassured that everything was going well at home.

CONCLUSION

Serving as part of Operation Enduring Freedom, as a member of CFC-A Coalition HQ, was a unique experience. I already knew that, as Canadians, we enjoy a fantastic quality of life. My stay in Afghanistan confirmed this in every way. I no longer see my minor daily problems in the same light, and I have trouble understanding how people can complain so much in our society. Since my return, I appreciate everything around me. And from a professional point of view, the experience could not have been more enriching. I was faced with highly unusual situations to which I had to adapt constantly. I can honestly say, the CF prepared me well to meet those challenges.

CHAPTER 10

FULL SPECTRUM LEADERSHIP CHALLENGES IN AFGHANISTAN

Colonel Bernd Horn

The three-vehicle convoy pulled away from the Task Force Afghanistan (TFA) headquarters building seemingly invisible to the normal hustle and bustle of people and vehicles scurrying about the inner confines of the Kandahar Airfield. However, those in the vehicles were keyed up – they were going outside the wire. Only a short time before the convoy commander had given orders to his section and passengers on the immediate action drills in the case of contact, as well as the location of the ammunition and extra weapons. There was no mistaking the charged atmosphere or the seriousness of what was about to transpire – to these troops it was clear that they were at war.

The convoy, consisting of a 17 ton light armoured vehicle (LAV III) armoured personnel carrier, an electronic counter-measures "G" Wagon jeep and a "Bison" Armoured Vehicle General Purpose (AVGP), emerged through the heavily guarded gate and moved into a loosely guarded Afghan National Army (ANA) controlled area. As the vehicles passed a giant berm on the side of the road, everyone cocked their weapon, loading a round into the chamber. Seconds out here – literally – meant the difference between life and death.

As the vehicles roared down the dirt road, the vicious dust kicked up by the lead vehicles lashed at exposed skin. The vehicle commander could be heard over the intercom preparing his soldiers for the next leg. "Okay, when we turn onto the main highway, you gotta stay alert," he coached, "watch your arcs, keep a sharp eye behind us." The convoy eased through the ANA checkpoint and entered the notorious Highway 4 that led to Kandahar City. Once again they would have to run the gauntlet of "IED alley."

As the vehicles turned onto the paved highway a large group of locals watched them from a dirt parking lot across the road. Some were standing in solitude, while others were in small groups. Others sat in cars. None seemed to be there for any tangible purpose. None seemed at all threatening. At least one individual was talking on a cell phone. On the surface it certainly had the appearance of nothing more than a local gathering spot. But then again, outside the wire nothing was what it seemed to be.

As soon as the tires gripped the asphalt the vehicles shot off at speed. The military vehicles led by the monstrous LAV III thundered down the middle of the road forcing oncoming traffic to drive on the opposite shoulder. Improvised explosive devices (IEDs) necessitated the aggressive driving style. Driving too close to the edge of the road increased the lethal effect of a roadside IED – and distance mattered. A few feet could make a dramatic difference.

As the sentries focused on the surrounding environment they were bombarded by a panoply of contrasts. The countryside was barren, desolate and harsh, yet held a strange beauty. Similarly, the sentiments of the local population reflected a startling array of contrasts in stance and bearing. The old men gave the convoy scant attention or ignored it outright as if it did not even exist. They seemed to embody a stoicism, which radiated a resiliency and patience that carried a nuance that this too would pass. The children, as always, added a carefree exuberance and would run in bunches towards the road and wave. Conversely, the young and middle-aged men would glare – their hostility and resentment barely concealable.

The soldiers' scrutiny, however, washed over the local population. As the convoy hurtled down the road the hyper-alert soldiers scanned the entire countryside, as well as the roadside for potential threats. The task was enormous. The ground was hilly, rugged and dotted by mud walled villages and huts that could hide hundreds of attackers. Throughout the fields peasants were gathered in plots of land that appeared far too barren to sustain any form of growth, yet they picked away at the earth anyways. Elsewhere shepherds watched over flocks of sheep, some individuals tended camels, while almost everywhere people just stood and watched or walked along the road. Spotting a potential belligerent while moving at speed was extremely difficult.

Roadside surveillance proved equally challenging. Debris and garbage littered the entire route. Scrap metal, old car hulks, piles of bricks and rubble or dirt, as well as garbage of every sort potentially hid a deadly IED. In addition, cars could be seen parked on the side of the highway. Some were in the process of repair while others were simply abandoned. And then, there was the traffic. Cars and trucks of all sizes and states of repair travelled in both directions. In such a saturated state of constant motion it was virtually impossible to differentiate friend from foe. In total, the threat environment was extreme, yet non-existent. What was a threat and what was the simple reality of existence in a destitute third world nation?

Within the context of this ambiguous, uncertain and lethal environment the convoy ploughed on. "Okay, heads-up," shouted the vehicle commander into the intercom, "this is where we got hit the other night." He was referring to the failed ambush two nights prior, where a similar convoy was attacked by an IED, which was luckily triggered too early by the insurgents and missed the vehicles. However, in that case the insurgents followed up with a volley of rocket-propelled grenades (RPGs) and machine gun fire, all just missing the vehicles as well.

As the convoy passed the spot, not a sound was made, yet the collective sigh of relief was tangible. The convoy then passed through a defilade, two hilly outcrops that dominated the road and provided would be assailants with concealment and cover. Close by were a number of villages that could easily absorb fleeing gunmen in a maze of anonymity. With eyes glued to the passing landscape a brief glimmer of hope radiated from the otherwise drab and dreary surroundings. In the doorway of a mud hut a woman completely covered in a *Burkhas* stood and waved enthusiastically at the passing convoy. Veterans who had conducted foot patrols in Kabul in previous missions had often recounted stories of passing women who without stopping, turning their heads, or bringing any form of attention to themselves, would say thank you to the soldiers as they passed. This simple gesture momentarily brought a smile to the faces of some. It was one small reinforcement that the operation was important to improving the lives of others.

Scanning ahead, the threat rich environment lay unabated. Slow "jingle trucks"[1] laboured up the road while another large 10-ton truck sat on the roadside as its occupants apparently attempted to strap down a loosening load. The countryside was again flush with villages, houses and activity. As the convoy pressed to the left to swiftly pass the slow moving vehicles the world became surreal. A large orange fireball erupted without warning and ballooned high into the air engulfing almost the entirety of the road. The explosion was strangely muted, with a barely audible "krrummpp." The fireball was quickly followed by thick black smoke and the concussion. A vehicle suicide bomber in a Toyota sedan laden with former Soviet bloc artillery ammunition rigged to a triggering device saw his opportunity and as the LAV III made its move to pass the slow moving vehicles, accelerated out from behind the truck. Once he was close to the middle of the passing armoured vehicle the suicide bomber detonated his deadly cargo.

The "G"-Wagon and Bison within 30-60 metres of the point of impact reacted instinctively as they were engulfed in the smoke and debris. Both vehicles swerved to the left around the blast and proceeded straight through the dense smoke. Once clear of the ambush site a cordon was quickly established. It took less than a minute to assess the situation. The suicide bomber's vehicle lay in two distinct but very small pieces of twisted metal across the road. The LAV had taken the blast virtually point blank but the driver was able to manoeuvre his vehicle safely out of the ambush site. All waited anxiously for the SITREP [situation report] to determine if there were any casualties. Miraculously, there was only one evident, but it was serious.

The LAV III had done its job well. Although it sustained heavy damage, it protected its crew. The serious injury was sustained by the crew commander, specifically, his arm that was exposed outside the turret closest to the actual blast.

As the security cordon was established the Bison went to the assistance of the casualty – members disembarked and quickly loaded the wounded soldier into the back where immediate first aid was applied and the injured soldier was rushed back to the airfield. Many were wide-eyed, this being the first time they had actually witnessed the carnage of war and witnessed first hand a maimed colleague lying before them bleeding, missing large chunks of flesh and moaning in agony. All reacted as they had been trained. If there had been any doubt in anyone's mind whether they were engaged in a conflict, it was quickly dispelled. In fact, this was only one of a number of serious incidents involving combat, deaths and casualties in a period of three days.

This in small part is some of the complexity our troops serving in Afghanistan must contend with on a daily basis. Every day, they must steel up the courage to face this complex, chaotic, ambiguous and lethal environment. As many of those who have served in Afghanistan will assert – when you go outside the wire, you never know for sure if you will be coming back. Clearly, the conflict that is being prosecuted in Afghanistan at present raises a large number of leadership challenges and issues at the tactical, operational and strategic levels. This following discussion is not meant to be comprehensive. Rather, it is an opening salvo to initiate dialogue and begin the intellectual process of attempting to assist our deployed troops in every possible way.

TACTICAL LEVEL CHALLENGES

At the tactical level, a number of leadership challenges exist in Afghanistan. Front line commanders will have to sustain morale, continually motivate their soldiers and maintain the fighting spirit in a complex and lethal environment. As with most conflicts and specifically in fourth generation warfare,[2] the "action" is more often than not spread out between long periods of boredom, tedious tasks and loneliness. With the current threat situation soldiers often find themselves in remote, austere and very harsh environments where such rudimentary niceties such as clean clothes, regular personal hygiene, fresh rations and an adequate water supply are difficult to attain. This is on top of a threat environment, where an attack through IED, mines or ambush could occur at any time.

Limited troops on the ground further exacerbate 48 hour rotations into patrol bases or back to main camps for maintenance and refit, prolonging the strain of the environmental and threat conditions. It is not uncommon for platoons to be deployed for stretches up to 25 days at a time in such locations for platoons.

Furthermore, the threat environment, as well as the cultural and language barriers, prevent troops from interacting with the local community on an open and daily basis. Sadly, the use of the local population as cover by insurgents has made identification of friend and foe one of the greatest difficulties. As such, a degree of distrust and alienation also feeds on the soldiers.

For those confined to camps, similar morale issues develop. The camps are relatively small, with limited activities, which breeds boredom. In addition, the dusty, dreary environment creates a monotonous almost depressing backdrop. Everything is a pall grey due to the dust, even the trees and vegetation that grow within the wire. Furthermore, for those within the camps, a higher level of fear is present as a result of their perception of the threat outside the wire and their ability to do anything about it.

To exacerbate these realities, when action occurs, it is most often a fleeting attack – an IED, suicide bomber or quick ambush. The attack occurs, creates havoc and leaves death, destruction and casualties in its wake – often without anyone seeing the enemy or firing a single shot in anger. The initiative often seems to lie almost entirely with the antagonists. They decide where, when, how and who to attack. As such, the coalition response is equally often purely reactive because it is unable to field the necessary resources to dominate the terrain. The inability to hit back can cause the soldiers to feel impotent. It can build frustration, fear and a sense of futility, if not hopelessness. When casualties occur, particularly deaths, morale is dealt another severe blow. As such, the leadership challenge to maintain the aim and mission focus, as well as the overall initiative in the campaign, is immense and critically important.

Coupled with the need to sustain morale, particularly in light of the factors aforementioned, is the leader's role in ensuring that soldiers continue to practice a healthy outlook in regards to the local population. It has long been recognized that culture is to insurgency what terrain is to conventional mechanized warfare. However, as already indicated, in the current environment it is sometimes difficult to breach the cultural barrier, particularly one that is so xenophobic. Moreover, it is not unusual for soldiers who are attacked to feel angry and betrayed. They deeply believe that they are serving in Afghanistan to create a better society for its people, yet, they are continually attacked by seemingly invisible antagonists who appear to operate effortlessly in the very Afghan society that the soldiers are trying to improve and protect. Although most understand that the average Afghani is just trying to survive, the resentment still builds with each attack, with every casualty and especially with every death.

The nature of the conflict fuels a spiral of antagonism. In essence, it is a vicious circle. As coalition forces continue to be targeted by IEDs and suicide bombers, they have no choice but to take the necessary actions to protect themselves. This, however, comes with a cost. As convoys drive aggressively down the centre of the road, they force local Afghan traffic to scurry for the shoulder. As they physically bump traffic out of the way, or threaten vehicles who follow too close by pointing weapons, or create collateral damage due to attacks against them and/or defensive or offensive operations – they risk alienating Afghan nationals. With every action taken against the population at large, regardless of justification or cause, a cost is incurred. Coalition actions could potentially push Afghans to support the Taliban, or at least cause them to turn a blind eye to Taliban activities. Yet, to do nothing and continue to be hit without taking some action – feeds soldier disillusionment and the potential to lose Canadian public support for the conflict, if it appears that the country's troops are put at risk without the ability to take the necessary steps to defend themselves. Moreover, if a safe and secure environment is not created for the local population, there is almost no hope of creating support for the new Afghan national government.

In the end, it is a delicate balancing act. Some form of action must be taken as the initiative cannot be left to reside with the insurgents. Afghan nationals, our soldiers themselves and particularly the Canadian public must all be made to feel that progress is being made on the 'front lines.' All want, and need, to see progress that justifies the sacrifice and expenditure in national blood and treasure.

This reality carries another key leadership challenge – making the integrated battlespace work. Successful counter-insurgencies are as much about political and economic issues and initiatives as they are about military action. Security is fundamental and indisputably a key to success. However, military action to ensure security is an uphill battle without the corollary political and economic progress. This means that tactical leaders and their subordinates must learn to communicate, cooperate and work with non-military personnel such as representatives from Foreign Affairs Canada (FAC), the Canadian International Development Agency (CIDA), the Royal Canadian Mounted Police (RCMP) and other government departments, as well as non-governmental departments (NGOs) on a daily basis.

This can sometimes be a greater challenge than it appears. These agencies have different agendas, alien organizational cultures and differing philosophies. The greatest problem is one of ignorance. None of the players fully understand who the other participants are; what they do; their mandates; or how they actual operate. Other government departments (OGD) and civilian agencies are normally not accustomed to military directness or command structures. In addition, ironically, they are most often nowhere near as flexible, more bureaucratic and more risk averse than the military. Where the military looks for quick solutions and immediate results, the developmental agencies focus on long-term sustainable development. Not surprisingly, timelines, approval mechanisms, communications and organizational methodologies all vary and require both patience and tolerance. The integrated approach unquestionably can create some leadership challenges that must be addressed. All military and civilian personnel must quickly be educated to the reality that success depends on an effective, cooperative military-civilian integrated approach to the counter-insurgency effort.

A fundamental element of this integrated approach for tactical leadership will be to also improve the working relationship with the host nation forces. After all, counter-insurgency operations are local actions and they are dependent on police and effective governance.[3] More importantly, in such a primitive tribal xenophobic culture as in Afghanistan, there must be an Afghan face to all solutions and governmental/coalition actions.

But this can create imposing challenges. First, the difference in language and culture can create a barrier, although this can normally be overcome with interpreters and sound cultural awareness training. More problematic is the matter of reliability, integrity and credibility of the Afghan governance structures and security apparatus. A myriad of concerns abound. Are the host nation army or police personnel loyal to the government and/or to the Coalition? Are the appointed governors, district chiefs and other governmental appointees loyal? In essence, who can really be trusted? Moreover, some elements of Pashtunwali (their cultural/honour code), as well as their innate tribalism, is alien, if not at times offensive, to Canadians. Such issues as the treatment of children and women, rampant corruption and the conduct of criminal activities pose serious challenges.

In addition, there is a the large amount of work that needs to be done to professionalize the ANSF, which will require the effective cooperation of all parties over a long period of time. Their often inadequate training and equipment make the ANSF effectiveness questionable – particularly if their loyalty is suspect. Overall, the Afghan National Army (ANA) is rated as fairly motivated, courageous and willing to fight – if they are supported directly by Canadian or Coalition troops. The inherent task for leaders is to build capacity within this institution so that they are both capable and willing to take on security tasks by themselves. The other problem is their shortage of numbers. This will be dealt with later.

The Afghan National Police (ANP) is a different problem. Young, untrained and answerable to the governor of their respective province, they resemble and behave more like common thugs than police. They wear no uniforms, and are notoriously corrupt, unreliable and untrustworthy. Complete detachments watching checkpoints or outposts are regularly found fast asleep and usually 'wasted' on hashish or marijuana. Significantly, large elements therein are suspected of being sympathetic to, if not in league with, the Taliban.

Faced with these challenges, it is often tempting for a commander to want to do the mission alone as this is in many ways far easier than trying to work with host nation forces. However, that does not build capacity. And, unless the Afghan government attains an ability to provide its own security, Coalition countries will be condemned to remain in Afghanistan. In addition, the use of ANSF provides additional manpower, a pool of individuals who know the language, culture, tribal customs and contacts, as well as terrain. Moreover, they put an Afghan face to an Afghan problem.

However, issues are still present, even with an "Afghan face." At present, the population does not identify with the national government. They fear the ANP, which is corrupt and feeds on them through brutality and extortion. They avoid the prevailing judiciary or governmental bureaucracy because they too are perceived as corrupt and bribes are necessary to get the least amount of satisfaction. For those reasons, as well as their strong tribal culture, 90 percent of their problems are solved through the informal process involving the local mullahs. These dynamics exacerbate the leadership challenges of the tactical leader.

Nonetheless, undeniably, the involvement of host nation forces is key to overall success of any counter-insurgency campaign. As such, developing close, interactive and useful relationships with all players in the new security environment will be an important initiative tactical leaders must pursue to ensure the success of the 3D (defence, diplomacy and development) approach, which is central to the overall success of the mission.

Yet another leadership challenge for tactical commanders will be to ensure all their personnel maintain the proper combat mentality. This has many facets. First, personnel must be capable of operating in an ambiguous, chaotic, volatile and rapidly changing battlespace. In addition, they must be able to think in non-traditional, non-western ways and think in terms of the enemy's perspective. Leaders and their followers must also be able to transition through the entire spectrum of conflict seamlessly – in essence they must be able to fight the "three block war." Simply put, military personnel must be able to transition from humanitarian operations, peace support or stability tasks to high intensity mid level combat, potentially all in the same day all in the same area of operations.[4]

In addition, importantly, leaders themselves must come to grips with, and help their subordinates to cope with, mass casualties and combat death. This is the single biggest shock and most difficult psychological challenge all participants are facing. Despite the fact that everyone acknowledged that there would be casualties and deaths in theatre, and they all witnessed ramp ceremonies on television before deploying, almost all were unprepared for it when it occurred. Leaders at all levels struggled with making sense of the deaths themselves and then had to try to comfort and make sense of it for their subordinates. "People are choking on a richness of experience," commented Padre Robert Lauder, "they are trying to metabolize it; trying to understand the new environment, but there is too much to chew, too much to swallow but they can't spit it out."

Another element of the proper combat mentality is the comprehension of levers that can influence and achieve military, economic and political objectives. In 4GW kinetic combat power is not always the most effective tool or weapon. Leaders must ensure that their personnel understand that money, medicine, fuel, food, access to education, employment opportunities, public works projects, respect and particularly information are all important enablers to achieving the mission. These non-kinetic, non-military tools are force multipliers that can dramatically change the threat picture and effectiveness of the insurgents.

A final aspect of maintaining the proper combat mentality is cultivating an understanding of the magnitude of the mission and the soldier's responsibility in making the operation a success. Repeatedly, commanders on the ground lament that not everyone in theatre, or at home, has fully grasped the scope of the conflict. This is not a peace support operation – rather it is a lethal battle zone. Although this problem is being quickly mitigated through a large amount of experienced personnel rotating through theatre and as a direct result of the lethal environment with its almost daily "troops in contact" (TICs) it is still an issue leaders must be conscious of.

The "do-gooder" polite Canadian attitude and trusting nature that inexperienced personnel arrive with must be contained. For example, during Rotation 1 in Kandahar when an Afghan national was caught in Kandahar Airfield taking pictures, one of the Canadian service personnel present tried to make a case for giving him back the camera since it was worth a lot of money. It totally escaped this individual that the perpetrator was spying, for whatever motive, with the end result of passing on information to the enemy. In a more serious incident, during the conduct of a Canadian civil military cooperation (CIMIC) meeting, the security escort failed to notice that their hosts had carefully removed their children from the meeting area. Once this was done, the CIMIC officer and his escort were attacked. They later stated they did not see it coming. Yet, had they not been so trusting, they may have noticed the change in setting – uneasiness of the elders, a shift of the crowd, whether positioning or make-up (e.g. the removal of children and/or women). There is a reason that the intelligence senior Sr NCO who provides an in-country threat brief to newly arrived personnel ends his 40 minute lecture with "Now its important to remember, they [Taliban] ARE trying to kill you."

However, part of the leader's challenge is also to ensure balance remains. Although it is extremely difficult to tell friend from foe, and the Taliban do utilize the population as a cloak to hide behind, leaders must ensure the combat mentality does not sink to the lowest denominator where individuals forget their responsibilities under the Law of Armed Conflict. It remains critical that everyone continue to try to minimize collateral damage and the endangering of the lives of innocent civilians.

In the end, the proper attitude and mindset are critical to reducing casualties and achieving mission success. This ensures leaders and soldiers conduct operations in an effective and focused manner, reducing risk where possible without compromising the mission. In the end, it helps guide everything they do – all their actions, towards a goal; towards achieving a specific end that assists in executing the campaign plan. The end, most often will not be a military one, but rather a political end that assists in creating a secure environment, building trust or nurturing support for the new national government. As such, leaders at the tactical level are critical to that success. To them falls the Herculean task of balancing the enunciated challenges and completing the business at hand.

OPERATIONAL LEVEL CHALLENGES

But the leadership challenges do not stop there. If the leaders at the tactical level are to be successful, then they need guidance and support from the operational level. It is at the operational level that strategic aims, or the ends to be achieved, are attained as the result of the skilful tactical employment of forces in accordance with a carefully designed and planned campaign plan. Operational level leaders understand that the campaign plan will extend both temporally and spatially well beyond what is experienced at the tactical level. These commanders must build in tolerance for ambiguity and concentrate everything on the operational end-state. Fundamentally, the campaign plan provides the operational constructs (i.e. ways) that are in turn reduced to tactical actions (means) to achieve the required objectives. As such, one of the greatest challenges for the operational leader is to ensure that the actions of his or her subordinate leaders contribute to the resolution of the conflict. Conducting patrols, raids, or sweep operations without tying them to the larger plan, political context, or to a specific goal to be achieved becomes in the end a hollow effort. In the end, the preference of many commanders for kinetic solutions is not always well suited for 4GW.

As such, the challenge for the operational leader begins with ensuring a comprehensive understanding by all of the nature of 4GW, or more specifically the nature of the insurgency that currently grips Afghanistan. Typically, the resolution of political and economic issues is far more critical to a successful outcome than purely military action. No amount of patrolling or sweeps through the mountains alone can solve the problem of providing useful employment for the unemployed and disenfranchised young men of the country, who represent a limitless pool of potential insurgents.

Therefore, an effective integrated approach is a key challenge that must be mastered. How does a leader ensure a diverse, multinational force is welded into a coherent organization all with a clear concept of the operational end state desired? This must then be extended to civilian counter-parts, who must be integrated into the team and into the decision making process in an accepting way and on an equal basis. What is often lost is the fact that invisible cultural barriers (i.e. divergent attitudes, beliefs and values, as well as methodologies and organizational practices) restrict true cooperation. Often we do not know what we do not know, and we assume our perception of the state of affairs is accurate and mutual, when in fact ground truth may be an entirely different reality. As such, an enormous challenge for operational commanders is to create a conducive environment in which planning, decision making and the conduct of activities is done in an integrated manner that allows for the necessary advancement of political and economic initiatives and reforms in a safe and secure environment.

This is always easier said than done! Nonetheless, securing the "hearts and minds" of the population is a slow and difficult process. Building trust and credibility always takes time. It takes hard work over a long period to accomplish; yet a single careless action can destroy all that has been achieved. However, success depends on the integrated approach and the operational level commander is key to providing the requisite support to other government departments (OGD) and civilian agencies to make it happen. In the end, the host nation population will base their support to the new government on the perception of improvement to the security environment and increases in their standard of living – e.g. employment, availability of food, water and electricity, education, responsive and responsible government.

Instrumental to success for the operational commander will be the challenge of separating the insurgents from the population. This may not always be an easy task. First, coalition forces must demonstrate themselves capable of providing a secure environment, yet, it is virtually impossible to stop all acts of violence, particularly by small groups of committed fanatics who operate among the people in a very closed and alien culture. Second, in the case of Afghanistan coalition forces, they represent a force committed to eradicating poppy fields, which unfortunately is a viable and profitable crop in the otherwise barren and hostile Afghani environment. Third, coalition forces, particularly as a result of defensive actions they have been forced to adopt, often appear as a force of occupation, or at a minimum as callous towards the average Afghani.

Thus, the challenge will be one of generating an atmosphere where people feel they have a stake in the system and become committed to making it work. A huge element is providing a safe and secure environment – the other part of the equation, as already alluded to, is fuelling economic growth and a strong capable national government. Aid dollars and assistance from OGDs, both Canadian and international, are of great importance, but will never be enough. Economic incentive and growth can only be sustained if they exist within a strong credible host nation political infrastructure. But, how does one incorporate poorly trained, poorly equipped, and at times suspect, host nation police or military forces? How does one operate within an environment where tribalism is accepted and corruption still practiced? How does one build cooperation and trust between the myriad of players (of diverse cultural, organizational and philosophical backgrounds) who are all essential to a successful outcome to the conflict?

The weak state of governance and the lack of credibility of the ANSF create enormous challenges for the operational (and strategic) commanders. Afghans are xenophobic and largely, although not always obvious about it, intolerant to *kaffirs* (foreigners). Therefore, putting an Afghan face to everything that is done is key. But, if the Afghan face is an ANP that extorts money at checkpoints, steals and bullies the people; and if the Afghan face is an appointed governor who was a previous warlord who continues to use the public office to enrich himself, or is tied to the drug trade; and if the Afghan face is locally appointed officials based on tribal connections who are corrupt, then the Coalition's actions and presence is seen as supporting a corrupt government incapable of providing security or governance. This brings us back to why the Taliban was created in the first place in the early 1990s. The challenge here is colossal, because many of the factors that drive it are beyond the immediate control of the operational (and/or strategic leader).

The operational challenges, however, extend beyond the necessity of galvanizing a holistic approach that ties military action to political and economic reform and advancement. The seemingly indiscriminate suicide attacks and more deliberate ambushes, as mentioned earlier, can easily create an image that the insurgents hold the initiative. As such, the operational commander needs to develop a plan that severs the belligerents from the population base, thus isolating the adversary from support. In addition, he/she must disrupt and destroy the enemy's command, control, communications and infrastructure. More importantly, the operational commander must ensure that Coalition forces always maintain the initiative, which, in counter-insurgency, is critical. If the belligerents are reacting to Coalition forces – then they no longer control the battlespace.

As a result, the operational commander must ensure the realization of the campaign plan, which provides the vehicle to achieving the strategic goals. But this can often be a difficult balancing act. The very indiscriminate and asymmetric nature of 4GW necessitates agility in thinking and the rapid and flexible conduct of operations, as well as decentralization and the reliance on initiative at the lowest tactical level. It is a small unit war most of the time. As such, subordinate commanders must be allowed the freedom to conduct operations based on circumstances as they arise. A culture of adaptability and agility of thought is key. Yet, as mentioned earlier, the operational commander must also ensure that the employment of tactical forces achieves specific ends, or objectives in accordance with the operational campaign plan.

To complicate this factor is the operational context. First, the commander rarely has enough enablers (e.g. aviation, fast air, surveillance suites, artillery, psychological operations) and those that do exist in theatre are national assets and controlled as such by the respective donor nation. As a result, priority of use and national caveats are not the commander's call. Furthermore, the environment is extremely complex and the commander simply does not own the battlespace. For an example, a commander may have to deal with up to 30 significant incidents a day, including a mass casualty; a catastrophic friendly fire incident minutes before H-Hr for a brigade offensive; five different special operations forces (SOF) groups all from different countries, all operating in his/her active area of operations (AO), all of whom report to different chains of command. These issues are exacerbated by national caveats on force utilization; insufficient enablers, all of which you do not own; hidden international agendas, Host Nation limitations; and domestic national imperatives, just to name a few. Added to this is an extremely complex operational environment where it is hard to determine friend from foe and the terrain is some of the most difficult possible in which to conduct military operations.

Needless to say, current CF doctrine of mission command, a philosophy that promotes decentralized decision-making, freedom and speed of action and initiative is instrumental in achieving this outcome. It entails three enduring tenets: the importance of understanding a superior commander's intent, a clear responsibility to fulfil that intent,[5] and timely decision-making. However, the highly politicized environment and the inherent risk associated with mission command creates a significant leadership challenge for commanders.[6] The theory is easy – it is the practice on the ground in an ambiguous, complex and lethal environment that creates the challenge.

In addition, the restrictions and limitations placed on the operational commander as a result of the coalition context is another huge challenge. As briefly mentioned earlier, leaders must cope with diverse, nation-centric and/or at times competing national interests in pursuit of the mission, which all must be dealt with through complex chains of command. From the political/development side this also means players must seek policy directives and/or authority prior to engaging. Moreover, national caveats on force employment provide further restrictions. Amazingly, it seems not all force providing countries are prepared to let their forces fight. This coalition factor is further complicated by cultural and organizational differences and the egos of personalities and especially commanders. The importance of building personal relations is key.

Finally, perhaps one of the most difficult challenges is balancing ground truth with domestic expectations. National agendas and Canadian expectations of progress and good news stories in regard to reconstruction and a better Afghanistan to justify the cost in Canadian blood and treasure often run counter to the reality on the ground. Leaders face pressures "to get on with it" while still wrestling with a very dangerous threat environment on the ground that does not always permit the necessary freedom of movement or collaboration required to prosecute development or reconstruction programs. In addition, ISAF resource limitations and foreign national caveats on NATO participating forces places an inordinate burden on Canadian resources to provide security in Kandahar Province an AO of approximately 60,000 sq km. Although all legitimate factors, it takes great leadership and morale courage to tell the king "he has no clothes."

STRATEGIC LEVEL CHALLENGES

The necessity for strong leadership at the operational level is seminal. After all, as already mentioned, it is at this level that strategic direction is translated into operational and tactical action. But leadership challenges do not stop there. They also exist at the strategic level. Strategic leaders need to be skilled at translating policy objectives into sensible strategic objectives. In addition, they must work effortlessly with a wide range of non-national organizations such as other militaries and international organizations like NATO, the UN and the European Union (EU).

However, no challenge is more daunting than getting and maintaining national support for involvement overseas, particularly when that involvement is perceived as being dangerous and of questionable value to the national interest. Unfortunately, the problem is exacerbated by an uninformed public who cling to erroneous perceptions of the role and capability of their armed forces, as well as emotional and naïve assessments of what is in the national interest. As such, a huge leadership challenge, particularly because of the traditional perception, if not expectation of the government, that senior military leaders do not enter in public debate, is how to educate Canadians in terms of CF involvement overseas.

Closely associated with the challenge of gaining and maintaining support for the mission is the question of sustaining the war effort, particularly since successful counter-insurgency campaigns take long term commitment. There is both an internal CF and external societal implication. Internally, operational tempo will exhaust personnel, particularly the support trades and specialists, as well as the combat arms, who find themselves rotating in and out of theatre on a less than desired frequency. In addition, the budgetary impact of sustained operations, particularly as scarce valuable equipment is destroyed, could create other tensions and challenges. Continued calls for additional money to support the overseas effort could also alienate taxpayers, particularly if the necessary initiatives highlighted above have not been undertaken, or if they are unsuccessful. However, no challenge will be as daunting as maintaining public support if casualties reach an unacceptable level as perceived by Canadians, who have already demonstrated a low threshold of acceptance of the mission.[7] In all, the leadership challenge of maintaining support and sustaining the war effort will be a daunting task for leaders at the strategic level.

Unfortunately, it is not the only difficult challenge at the strategic level. Equally formidable is the requirement to comprehensively adopt the 3D approach at the national level. Most often, necessity and personalities on the ground in theatre make local arrangements work with varying degrees of success. At the strategic level bureaucracy, risk aversion, organizational dynamics and operating procedures (e.g. difficulty in finding personnel willing to deploy to dangerous theatres, rotation of staff, divergent philosophies on methods or outcomes, attitudes, stereo-types, miscommunication and/or incompatibility of communications, competition for government attention and budgets, personalities and egos) create difficulties. However, as already indicated, success is elusive without an integrated approach. Therefore, although the Department of National Defence is only one of many government departments responsible for executing government policy, its physical presence on the ground and in harm's way requires that its strategic leadership tackle the issue of attaining, nurturing and progressing the buy-in and support of such OGDs as Foreign Affairs Canada, CIDA and other departments and agencies necessary for the success of operations in Afghanistan. Moreover, they must attempt to encourage if not cultivate a bureaucratic dynamism within these organizations so that they become as flexible and capable of rapid deployment and action as is the military so that operations can be planned and executed in a fully integrated manner from the strategic, right down to the tactical level.

CONCLUSION

This chapter is in no way meant to represent a comprehensive examination of the leadership challenges that exist in Afghanistan. It is but an initial survey of some of the complex and inter-related issues that face our leaders at the tactical, operational and strategic levels. It also forms the foundation of continued discussion and research.

ENDNOTES

1   Jargon for 5-10 ton commercial stake trucks that are elaborately decorated with artwork, design and small bells – hence "jingle" trucks.

2   Fourth generation warfare (4GW) refers to a nonlinear, asymmetric approach to war in which agility, decentralization and initiative are instrumental to success. Furthermore,

4GW departs radically from the traditional model in which the conduct of war was the monopoly of states. In 4GW, non-state actors such as Hamas, al-Qaeda and the Taliban become serious opponents, capable of operations outside of their traditional areas of operation. Moreover, their definition of combatants diverges significantly from the traditional laws of armed conflict. According to military strategist William S. Lind, "fourth generation warfare seems likely to be widely dispersed and largely undefined … It will be nonlinear, possibly to the point of having no definable battlefields or fronts. The distinction between "civilian" and "military" may disappear." "The Changing Face of War: Into the Fourth Generation," *Marine Corps Gazette*, October 1989, 22-26.

3   Important to remember is that political legitimacy is measured in local terms and not Western constructs.

4   The concept of the "Three Block War" was originally articulated by former Marine Corps Commandant, General Charles Krulak who described it as an operational contingency in which soldiers would have to conduct operations spanning humanitarian assistance to peacekeeping and/or mid-intensity combat all in the same day and all within three city blocks.

5   The Commander's intent is the commander's personal expression of why an operation is being conducted and what he hopes to achieve. It is a clear and concise statement of the desired end-state and acceptable risk. Its strength is the fact that it allows subordinates to exercise initiative in the absence of orders, or when unexpected opportunities arise, or when the original concept of operations no longer applies.

6   This is exacerbated when one considers the "CNN effect," normally understood to be the omnipresent, 24/7, real-time news coverage that captures virtually any event at any time and instantaneously beams it to millions of viewers around the world. Its impact has created an enormous challenge for military personnel. This has led to the phenomenon of the "strategic corporal," where the tactical decisions or errors made by junior members can become strategic issues as they are beamed across the globe in real time and influence or incite negative and often violent reactions.

7   A poll conducted by Decima Research in April 2006 demonstrated that 46 percent of Canadians believed the mission in Afghanistan is a bad idea. CTV news, "Afghan Mission in Canada's Interest: O'Connor," http://sympaticomsn.ctv.ca/servlet/articlenews/story/CTVNews/20060410/afghanistan_debate_06_accessed 28 April 2006.

CHAPTER 11

REFLECTIONS ON AFGHANISTAN: COMMANDING TASK FORCE ORION

Lieutenant-Colonel Ian Hope

Under the baking Afghan sun we are rediscovering, by way of pain, that the first determinants in war are human. In combat, the power of personality, intellect and intuition, determination, and trust, outweigh the power of technology, and everything else. This stark reminder comes after 30 years of the Canadian Army following obediently the lead of our allies in combat development and falling victim to the seduction of the microchip. We have all come to see warfare as a clash of technologies, conveniently named by progressive thinkers *Third Wave, Network Centric*, or *Fourth Generation*. The high levels of journalistic prose, the mind-numbing powerpoint and jingoism, and the elevation of far-fetched assumptions to the level of indisputable truths have conspired to let us forget that war remains, now as it was when Agamemnon met Priam, a violent clash of independent human wills. At the moment of close-quarter contact between these opposing wills, when extreme violence, casualties, heat, thirst, uncertainty, and pressure that comes from the bottom-up (soldiers caught in situations beyond the scope of their experience and problem-solving skill looking to a commander in a manner that begs: "what the hell do we do now sir?"), tactical solutions and success come from the human heart and head. It is in preparation for this moment that young leaders must dedicate their energy, physical and intellectual, and develop their personal value system. They must come to trust themselves, trust in the heart and head of each other, and trust in the heart and head of their superiors if they are to survive and win this violent clash.

I learned this in command of Task Force (TF) Orion, the name I gave to the 1st Battalion, Princess Patricia's Canadian Light Infantry (PPCLI) Battle Group (BG) and its attachments: a Tactical Unmanned Aerial Vehicle (TUAV) troop, an HSS company, a Forward Support Group (FSG), and the Kandahar Provincial Reconstruction Team (PRT). I chose Orion to give everyone in this uncommon grouping of soldiers, sailors, airmen and airwomen a common identifier, something that might help them bond together more easily.

CHAPTER 11

Task Force Orion was created in September 2005 in Wainwright, at the beginning of pre-deployment training for our mission to Afghanistan. I had been in command of 1 PPCLI for 12 months. I had been warned of our mission to Afghanistan in October of 2004, and was personally told by General Rick Hillier, the Chief of the Defence Staff (CDS) in March 2005 that we would go to Kandahar under the American led Operation Enduring Freedom (OEF). So by September my RSM, Chief Warrant Officer Randy Northrup, and I had created a fairly cohesive battalion team that was focusing on the upcoming mission.

At that point, however, we needed to expand the "team" (our unit) to include the other components and attachments that would make up our BG. We did this by co-locating all of our people in one bivouac, by decentralizing combat support assets and embedding them within combat subunits, by providing training scenarios that concentrated on establishing broad trust in each arm and corps represented in the task force – mainly through extensive live-fire training – and by forcing people away from sub-unit identifiers (flags and colours) toward acceptance of a common – neutral – identifying symbol. I chose Orion from the constellation – representing the mythical Greek hunter of mountain beasts – that I knew blessed Afghan skies, so that our soldiers might look up and, seeing it, feel part of a larger entity, enduring and meaningful. I used this symbol deliberately to help create cohesion within this ad hoc grouping. It is in efforts to create and maintain cohesion that trust becomes meaningful beyond the boundaries of the individual soldier.

THE MISSION AND THE CONCEPT OF OPERATIONS

We deployed to Kandahar in mid-January 2006 and began operations there in early February. Our mission was stated as: "TF ORION will assist Afghans in the establishment of good governance, security and stability, and reconstruction in the province of Kandahar during Operation (OP) ARCHER Rotation (Roto) 1 in order to help extend the legitimacy and credibility of the Government of Afghanistan throughout the Islamic Republic of Afghanistan and at the same time help to establish conditions necessary for NATO Stage 3 expansion."

Ours was a transition mission. Canada had accepted the lead role in Regional Command – South (RC-S) for the period that would see the transformation of mandates from the OEF to the NATO-led International Security Assistance Force (ISAF). ISAF had expanded in 2005 from its presence in Kabul to take charge of northern Afghanistan (Stage 1) and then western Afghanistan (Stage 2). Stage 3 expansion foresaw ISAF taking over southern Afghanistan – the most difficult transition given the volatile nature of this region. It was thought that once Stage 3 occurred, then Stage 4 (eastern Afghanistan) could come about rapidly, unifying all of Afghanistan under one military headquarters (HQ).

No other nation had volunteered to conduct Stage 3 transition. It was impossible for the United States to do this effectively due to their existing commitments. The British did not have the capacity at the time, and the Dutch simply refused. Canada accepted the leadership role for both OEF and NATO at this critical juncture.

Our tasks were multifarious, divided into three broad categories: governance, security and reconstruction. These became our lines of operation (broad sets of connected objectives and tasks that contribute to the elimination of the key problems or obstacles to our mission). The key problem I saw was the winning and maintenance of the confidence of the people (who were at the same time subjected to considerable pressure by the Taliban to support their cause). In this sense, the struggle in Kandahar was essentially a clash of human will, between forces loyal to the government of Afghanistan and those opposing this authority, with the people of Kandahar in the middle being pulled, enticed, persuaded and coerced to chose sides.

The confidence and support of the people was essential to establishing freedom of movement and action for each of the opposing forces. We both relied upon the same source for such essential freedom and support. The people constituted a common "centre of gravity" over which we and the Taliban wrestled.

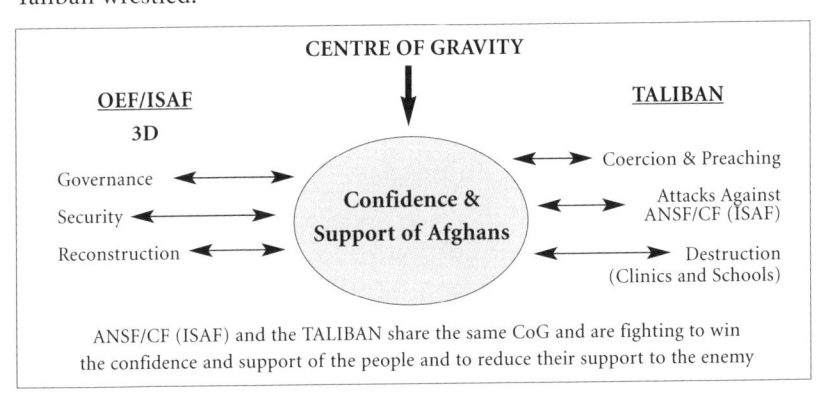

CENTRE OF GRAVITY

OEF/ISAF		TALIBAN
3D		
		← → Coercion & Preaching
Governance ← →	Confidence &	← → Attacks Against
Security ← →	Support of Afghans	ANSF/CF (ISAF)
Reconstruction ← →		← → Destruction (Clinics and Schools)

ANSF/CF (ISAF) and the TALIBAN share the same CoG and are fighting to win the confidence and support of the people and to reduce their support to the enemy

When you come to understand the conflict in these terms, you can easily see the importance of trust beyond institutional horizons. Establishing trusting relationships with local authorities and with the people was essential to swaying the centre of gravity to our side. This model is essential to understanding the nature of counter-insurgency operations.

As an operating concept, I intended that the 1 PPCLI BG become an extension of the PRT so that I could coordinate and synchronize all Task Force tasks in a unified effort under one chain of command. This centralized control was to be exercised in a manner responsive to desires of Afghan authorities and their people. To this end we established a combined Afghan/Coalition Provincial Coordination Centre (PCC) at the Governor's Palace, to help build the capacity of provincial civil, police and army leaders in managing security. Likewise, I wanted the PRT to operate downtown in order to work with the provincial staff to coordinate international agency reconstruction efforts and to assist in the functioning of the Provincial Development Committee (PDC). This would give the provincial leadership powerful centralized organizations that could assist them in exercising their authority while building their capacity to govern effectively. While command of the TF was centralized and responsive to the provincial authorities, the execution of the governance, security and reconstruction tasks was very decentralized. I pushed elements of the manoeuvre sub-units out into independent sub-unit forward operating bases, located in areas that would allow them to work with Afghan National Security Forces (ANSF) and district leaders to improve governance, security and socio-economic conditions in key districts of the province. We organized in a very decentralized manner. We evolved the decentralized concept to the point where within a platoon there were multiple attachments.

To enhance provincial and district governance I wanted the TF elements to engage village, district, and provincial leaders to convince them to implement Government of Afghanistan (GOA) initiatives. I wanted every patrol to stop in villages and conduct local shuras (meetings) to hear out the people and to encourage the elders to engage district leaders. The patrols reported results to the officer commanding (OC) the respective sub-unit who in turn engaged district leaders to properly represent the villages to their provincial leaders. The PRT Commander and myself engaged the governor and his staff and provincial council to hear representation from the districts. TF patrols were also tasked with reporting village and district reconstruction requirements to the PRT and to monitor the progress of reconstruction projects in rural areas. The PRT led in assisting provincial leaders in establishing reconstruction priorities, building their capacity to manage reconstruction projects in a legitimate manner. At the same time, TF elements were to operate with and train the ANSF throughout the province in order to create ANA capacity to conduct independent COIN operations, and to build ANP capacity to enforce the rule of law.

While we were committed to this 3D (Defence, Diplomacy and Development) approach, two things combined to focus our efforts upon the security line of operation. The first was a loss of momentum in PRT activity resulting from the tragic death of diplomat Glynn Barry in January and the long review of the PRT by Foreign Affairs Canada (FAC) and the Canadian International Development Agency (CIDA) that followed with a cessation in their funding to the PRT for six months. The second was a steady increase in Taliban activity until June when it became obvious that a full-blown Taliban offensive was underway in southern Afghanistan that threatened our ability to achieve Stage 3. Our efforts were then drawn more and more into responding to this threat by executing a long series of sub-unit and BG operations to find, fix, and destroy the enemy.

TRUSTING OURSELVES – FINDING THE ENEMY

In attempting to find Taliban fighters, Task Force Orion was supported by a myriad of "enablers" (Electronic Warfare (EW) and imagery platforms, and various human intelligence (HUMINT) networks). However, we could not trust in these. The brigade did not have a sufficient number of enablers to cover every task, and our operations were normally of lower priority than any of the time-sensitive HVT (high-valued target) hunts being conducted by special forces. Therefore, even when we were allocated several enablers, they were often stripped away very early in an operation, leaving us relatively deaf and blind and with sinking feelings as they departed each time. When they were present, they provided useful intelligence and support, and made us feel more secure. But they were also plagued with the friction normal in combat: incompatibility of technologies; air gaps; low fidelity sensors; receivers that were not rugged enough for combat; vehicle and equipment breakdown; bad atmospherics and foul weather; restrictions on intelligence sharing that are inherent in coalition operations; and human error. Friction degraded our sensor and communications networks to the point where they were incapable of producing more than a fleeting glimpse of a portion of the enemy's force; temporary visibility upon only one piece of an immense and dynamic jigsaw puzzle.

To compensate for lack of effective ISTAR (intelligence, surveillance, target acquisition, reconnaissance) we began to disperse across the province and live amongst the villagers in order to develop local trusts and gather intelligence through human contact. This forced the soldiers to live in the light armoured vehicles (LAVs), or beside the Gun Lines, often for three or more weeks at a time without shower or proper toilet, in hot, dusty, arid conditions. We lived this way beside the Afghan people, in the face of the enemy, with good psychological impact. Communication intercept (ICOM) and other EW sources often confirmed that each of our platoon, sub-unit and BG manoeuvres was causing a disruptive affect on Taliban efforts to mass in Northern Kandahar. I began to sense in March and April that, while we were seldom making direct contact with Taliban forces, we were being successful in keeping the initiative.

The maintenance of operational initiative became something tangible to me; I, and the subordinate commanders, could sense when we had the enemy confused, or when they seemed ready to act. I began to trust intuition, especially if it complimented local HUMINT (provided by villagers or ANSF). Our concept of operations evolved to where we continuously planned and executed multiple large manoeuvre operations in order to find, fix and finish individual enemy groups in an extended effort to keep the much-valued operational initiative. Reflecting upon this now, I realize that this dynamic was seldom if ever really replicated in our peacetime training, where the ebb and flow of the initiative was only academic (probably because such things never enter a Master Events List).

In exercising the *sense* function I relied less and less on an inventory of ISTAR sensors and more and more upon personal *sense* of where the enemy was, what he intended to do, and how he could be deceived. TF leaders personally reconnoitred districts and analyzed local HUMINT sources to identify trends and to guess if an enemy was in a general area (almost always this was relayed to us as a 20-40 man Taliban group hiding in the vicinity of a particular village). We then attempted to manoeuvre into that district quietly under cover of darkness, using deception, and – as much as possible – isolate the village by using thin blocking and cut off forces. We would conduct manoeuvre (cordon and searches) and fires (show of force artillery or 25mm fire) to produce enemy ICOM chatter, and from this ICOM chatter (or HUMINT from local nationals), we would attempt to vector in upon the enemy's locations. However, in the end, finding him was almost always a result of advance-to-contact in the close country where he hid, and was confirmed by the exchange of fire at close quarters.

Intuition and trust were central in our efforts to find Taliban forces. Over time we began to decipher which Afghan information sources were reliable (and which were not) and to trust our instinct, regardless of whether or not they were confirmed by our own ISTAR capabilities. I began to trust the subordinate commanders to act – or not – upon this local HUMINT. Given the lack of "actionable intelligence" about the enemy, there were few specific tasks given to TF Orion by the brigade, so it was up to us to establish tempo and task. Operational tempo and effectiveness were almost entirely dependent upon the personalities of the commanders. We could have very easily hunkered down into localized routines in relatively safe areas, following a pattern of "framework patrolling" and by this be seen as "busy", when in fact we would achieve little (except to surrender the initiative and set ourselves up for attacks by improvised explosive devices (IED)). Instead, I forced myself and encouraged – and trusted – the company commanders to be perpetually proactive in hunting the enemy, forcing everyone to live "outside the wire" and to continuously manoeuvre and to engage locals. This was difficult for everyone, especially after casualties were incurred, when there was always a natural tendency to adopt caution and reduce tempo. It was at these times especially that I trusted the subordinate commanders to demonstrate an aggressive spirit and maintain operational initiative through offensive action. Not all people are equal in this regard and the Army must learn that robustness and determination are not nice-to-have qualities in manoeuvre commanders; they are core qualities, and we must work hard to select the few who have the tenacity to sustain a high level of resolve in the face of danger.

TRUSTING EACH OTHER – FIXING AND FINISHING THE ENEMY

Once found, we would attempt to fix the enemy with fires and destroy him in close-quarter combat. Neither of these things was easy. It required us to stay within 100-150 meters of the enemy and to coordinate fires before physically moving to clear his positions. Doing so was contrary to human nature. Like operational tempo, the degree of success we had in close quarter combat became personality-dependent. The majority of soldiers when fired upon for the first time would seek to disengage back toward the "last safe place" they occupied. After several encounters they repressed this urge but would be very reluctant to advance in contact (especially when separated from their LAVs). Forward movement, or staying in place, on the close-quarter battlefield (especially after night-fall) depended upon the continued presence of battalion and company commanders, supported by the "natural fighters" in our ranks. It became evident to me in May that the number of true fighters we had was a small minority. By fighter, I mean those men and women predisposed to keep fighting regardless of violence and danger; those who repressed fear not just because they wanted to remain with their primary group, but because of an overwhelming desire to beat the enemy; those who truly wanted to hunt the enemy and make him the victim. I would estimate that there were only 6 or 7 such individuals in every forty-man platoon. Yet, their stalwartness almost always became the psychological pivot point for the action of a section, platoon or company engaged in intensive battle. I came to rely upon the courage of commanders and this small number of fighters in each platoon and company to override the inherent fear of close-quarter battle and to ensure that we kept the enemy fixed before closing to destroy him. Trust – not authority – was the essential element.

It became silently understood that trust and cohesion required an equitable sharing of risk. I travelled everywhere by vehicle, and slept on the ground, and avoided using helicopters because of the impression this left upon the troops. Instead I would personally lead all road moves until the point where I felt we would begin a final advance to contact. It is my philosophy that the commanding officer must command the lead vehicle, must navigate and make the tactical decisions during all phases of movement. This served several purposes; it alleviated a subordinate commander of an additional task of route navigation prior to a battle (allowing him to focus on the close fight after he was delivered to it); it took advantage of the most experienced crew in the Battle Group – my 9er Tac traveled over 9,700 km during the tour and knew the routes and terrain better than any other crew; and this helped avoid unnecessary navigation errors and personal embarrassments to junior leaders who needed their confidence intact before going into battle; and it demonstrated a sharing of risk. The approach came with cost; 9er Tac LAV

IN HARM'S WAY
CHAPTER 11

was hit four times by IEDs and Rocket Propelled Grenades (RPGs), resulting in the loss of two of my crew, Master-Corporals Loewen and White, grievously injured on separate occasions by shrapnel. This demonstration of shared risk went far beyond any authority granted by rank and appointment to promote trusts essential to the fighting spirit of Task Force Orion.

The company commanders were also always seen well forward in the fighting. They were entrusted to personally lead the close-quarter fight, and to assist me in my role of coordinator of the close battle. During firefights, my job was to work with each sub-unit commander to resource and coordinate fire and manoeuvre, apply air support and intelligence support, exercise liaison with ANSF, and control the evacuation of casualties and prisoners. Exercising direct control over the coordination of supporting functions allowed the subordinate commanders to get on with the harsh business of close-quarter fighting. I trusted and relied upon these commanders to apply the determination, fierceness, and courage that this hard task required. In turn, the junior leaders could rely upon their superiors to provide the best coordination of support. Task Force Orion applied many dozens of support fires – sometimes at danger close – without blue-on-blue casualties because of the active work of all commanders in controlling the fight.

In the coordinating function, I found that the evacuation of our dead and wounded, and the recovery of our damaged LAVs to be of the highest importance and I was prepared to stop all forward advance in order to get these things correct. This was different than what we had taught and practiced in training; where we preached that we would not let casualties slow down forward momentum, having second in commands (2Ics) and company sergeants-major sort casualties out while commanders continued to attack. The reality of wounding and death changed that notion. Commanders became personally involved in evacuation. It became our number one concern in battle, with positive effect. Soldiers came to understand that if wounded or killed, their recovery from those hellish close-quarters was the first priority of the chain of command, beyond any "prize" of Taliban defeat. We did not have to articulate this, it came naturally in mutual understanding and, I believe, in trust.

Finishing the enemy in close-combat meant killing or capturing them, and then capitalizing upon this act with aggressive information operations intended to reduce enemy confidence and raise that of the local populace (and our citizens back home). We never construed the use of lethal force as a negative action, or a "last resort". By the time that Task Force Orion came to be engaging enemy forces on a daily basis (in June, July and August), I had stopped thinking about clever ways of winning without fighting. I realized that "shattering enemy cohesion" and maintaining the initiative required a degree of physical destruction. In the emerging discussion of Effects Based Operations I contribute this; the best second-order and third-order effects are produced by first killing some enemy and then using such destruction to best advantage in information operations. In cultures where degrees of violence are largely accepted and indicative of strength, this is how insurgents are beaten. The idea of winning without hard fighting is complete folly. However, the act of destruction only served to finish an enemy group if we ended the action by remaining on the battlefield – forcing him to withdraw – and then beating the enemy to the punch by communicating with the Afghans first, telling them about our successful reduction of enemy fighting power and especially about how the enemy withdrew from the area, conceding our superiority of arms. By this we achieved local (tactical) psychological advantage that allowed us to maintain the operational initiative and to begin to consider re-introducing governance and reconstruction projects.

In a similar vein, I never left a broken fighting vehicle on a battlefield (never mind a wounded or dead soldier) for the enemy to gloat over in the international media. In fact, in operations where we had fighting vehicles destroyed, I deliberately stopped our efforts to fix and destroy a particular enemy group in order to recover our downed LAV before the enemy could photograph it and from that gain a psychological victory in our misfortune. We would lose more from the image of a burning LAV on Al Jazerra than we could possibly gain by the infliction of casualties on yet another small enemy group. Physical reductions in counter-insurgency are less important than psychological reductions by publicly attacking the enemy's status. But you require both to win. The true finishing of the enemy was achieved each time we reduced the status of a Taliban group to "a has-been" in the eyes of the locals.

The idea that the commanders of Task Force Orion wanted the soldiers of the Task Force to fight, and by destruction to win, was fundamental to troop morale. Over time, the soldiers came to know that their commanders were constantly looking for enemy for them to kill or capture, and had trust in their ability to do just that. During the period April-May, "A" Company had the misfortune of operating in the mountains of Showali-kot, where the enemy avoided contact, but made maximum use of IEDs. Morale in "A" Coy was affected. To reverse this, and to demonstrate trust in the company, I ordered them into the fight in Zharie/Panjwayi and Helmand in July, where they demonstrated their proficiency in arms, and where they exacted retribution upon the Taliban. Their morale soared.

MISSION COMMAND

Task Force Orion operated under the principles of mission command. Most often the companies operated independently, with 9er Tac moving between them fulfilling coordination functions. Frequently, companies would fight disparate engagements. At the point of highest tempo in July, the Task Force fought simultaneously in dispersed operations in Panjwayi, Sangin, and Hyderabad, then Panjwayi, Nawa and Garmser (each company fighting 50-100 kilometres from the others). Superb sub-unit leadership, supported by outstanding operational coordination by a brilliant battle group staff, handled all these separate fights. At no point in this period of prolonged dispersed combat did I have knowledge of the tactical situation in each sub-unit fight, nor did I need it; because I trusted each sub-unit to continue to execute our concept of offensive manoeuvre operations designed to reduce enemy capacity, maintain the initiative, and continuously create conditions that I could exploit positively in information operations.

This is not to say that I did not expect mistakes to be made. All the commanders of Task Force Orion made mistakes in operations; mostly because the experience of sustained combat operations was new to us and we had no paradigm or experiential guides. Tactical errors were made that cost life and limb. We who were responsible for these decisions and who gave these orders had to deal with that knowledge immediately – in the face of continued operational pressures – without compromising the faith and trust necessary for mission command to work. Task Force Orion instituted learning processes to capture lessons learned after each major engagement and to disseminate these throughout the AO. For my part, I emphasized use of the after action review (AAR) at all levels, sometimes running them myself, in the field, after the fighting was over.

In all of this we did not dwell upon mistakes. Sometimes we did not even speak of them. Nothing is more damaging to trust and self-confidence than to have a man who had acted decisively in combat, but had been a victim of bad luck, to be interrogated by a commander who was not present and had not felt the pressures and complexities of that moment. While I desired that commanders determine quickly what had gone right and what had gone wrong, I did not want this to create an environment of blame. Dealing with loss was a powerful enough motivator to quick learning. Finding quick forgiveness of one another went a long way to avoiding the trap of blame games.

DISCIPLINE, TRUST AND COHESION

Mission command was sustained because trust had been established well before the fighting started, with a unit culture that promoted self-criticism and self-adjustment. I deliberately steered away from the Canadian Forces (CF) tendency to order investigations. When we were asked our opinion of participating in standard investigations into all combat actions, I baulked at the idea for two reasons. Firstly, no investigative group from Kandahar Airfield or from Canada was qualified to understand or replicate the circumstances of the combat we faced. Secondly, an automatic default to outside investigations was an automatic statement of distrust in commanders who for the first time in half a century were dealing with the unknowns of combat. My response to the initial attempts by the CF to investigate every firefight was simple: if we look to blame someone for any of this, let it be first understood that when a man is wounded or killed on a battlefield, there is a lot of blame to go around. War is messy, confusing, and unpredictable. Predictability and certainty are never achieved. Determining cause and effect is extremely difficult if not entirely impossible. To attempt it as a bureaucratic standard would seriously erode trust.

I am not suggesting in these statements that we tolerated breaches in the Code of Service Discipline or the Law of Armed Conflict. If I sensed that these were at issue, there was only one recourse – swift and decisive action. Discipline is an absolute essential in combat units and clear boundaries must be drawn, expectations must be announced and disseminated, and transgressions must be firmly dealt with. This extended down to the seemingly trivialities of dress. I allowed soldiers to wear whatever load-carrying apparel, boots, and gloves they wished. I demanded that they wear proper personal protective equipment (PPE) and IFF (identification friend or foe), and took swift action when they did not. I would also not tolerate the proliferation of platoon or company badges or flags, as they jeopardize proper PPE and IFF, and because they reinforce sub-unit cohesion at the expense of unit cohesion.

It was beyond unit boundaries that things became more problematic. The investigative culture of the garrison CF was antipathetic to our work. We received orders to pass to the Military Police responsibility for ALL investigations for accidental and negligent discharges. This delayed the expediency of summary trial and left soldiers anguishing unnecessarily. It also served to diminish the authority of the chain of command. At other times, CF-sponsored investigations were conducted that I was not made aware of; these were synonymous to votes of non-confidence. These things breached the principles of mission command and of a trusting command climate. To help reduce the negative impacts and to reinforce the role and trust of commanders, I instituted a policy wherein any discharge of weapons outside of firefights – from either negligent discharges or escalation of force – required the company commanders to inform me in writing within 24 hours of the who, what, where, and when of the incident and to state whether or not they believed the incident required investigation for possible breach of the rules of engagement (ROE). I gave consideration to these and would act appropriately. As this policy required automatic reaction and direct involvement of company commanders, it served to reinforce their role, and to make us accountable for knowledge of subordinate actions. In this manner, it to a small degree helped reduce the garrison investigative mentality.

TRUST, FEAR AND LOSS

Risk of injury or death was shared, and I, and the sub-unit commanders, came to know intimately the fears of moving across that imaginary line that separates what constituted safe ground and what we knew to be a dangerous place. We came to understand why soldiers kissed their personal talismans before a fight, the fatalistic acceptance of death, how to deal with that visiting shadow that popped the dreaded question in my head "will I ever see my children again"; and how to dismiss fears and carry on with our tasks. These psychological oscillations are never present in peacetime training; they were the constant companions of those who lived outside the wire, and marked them permanently from those whose relative safety precluded having to deal with that dark shadow. What most visiting officers failed to grasp when talking with our soldiers was that the soldiers had begun to judge their leaders by how much personal risk they were prepared to assume. Helicopter "drive by" visits by senior leadership who left to go and sleep in comfortable places never left any impression upon the soldiers. The commanders left in the field began to understand that to have these soldiers respond right in a fight meant being present – up front – and demonstrating that rank made no one immune to the awful consequences of combat. Shared risk is essential in establishing hierarchical trusts.

To sustain a unit in prolonged combat there must be a great deal of hierarchical trust. That trust demands understanding and forgiveness; especially when it is clear that fear is visiting. I did not question any request by a subordinate commander who asked for a long water break, or who needed time to sort out communications issues before or during battles. Sometimes I, or one of my subordinate commanders, needed a moment to thrash the dark angel that had chosen that moment to visit, before picking up and carrying on. I knew that commanders were sometimes dealing with battles within themselves, but I never asked the embarrassing questions. To let a man know face-to-face that you see fear in him is to forever compromise him to you. To let their subordinates know that you have seen this fear in their commander would spread the fear far. Better to grant him time to deal with it and make it clear that you expect action consistent with the maintenance of the initiative over the enemy. Over time, most of us became familiar with the violence and uncertainty to a point where right-brain control (rational thinking) combined with fierce determination could win over fear. But it takes time and exposure to come to this point; and trust, understanding and patience are applicable in learning to fight.

If a soldier or leader was known to many to have faltered in combat from fear, there was only one recourse. The commanding officer, if he has any trust left at all, must place these individuals beside him, and prove to the collective that they are being granted a chance to redeem. They will not let him down.

Task Force Orion suffered 10 percent casualties over the course of six and a half months. Loss became familiar. Dealing with loss became a leadership challenge. The Canadian military has had a fragile nature regarding loss of life. A casualty during a peacekeeping operation would trigger widespread grief within units, and regimental history and memory focused upon dates and places of individual loss – even when attributable to accident. Raising the profile of loss to near-cult status had created a culture hard-pressed to deal with reality of sustained combat. We had to teach ourselves to be harder, to grant the moment for the fallen, but to move on rapidly. Wallowing would not do.

Difficult as this was for the soldier, it was harder for the commander. While the privates would mull over their two or three dead friends, I had to deal with them all. It took a great deal of rational thinking (studying the context within which each death occurred) and determination to handle this and to move on; especially if one cares deeply for each life and limb. It is in dealing with loss that Field Marshall Wavell's assertion takes full meaning; the most essential quality of command in war is robustness, a solidity of mind and spirit to the shocks of war and the grief of loss. This must become well understood to commanders entering theatres of war.

CONCLUSION

In garrison in Canada, units can go on for long periods of time without personal supervision of a commanding officer. Our administrative procedures and standard training scenarios make the role of commanding officer, and OC, relatively easy. I felt that I could give broad guidance and leave my battalion in Edmonton for months and feel confident that the unit would carry on quite well in my absence. Even in pre-deployment training, the scenarios presented were never so bad that the soldiers and company leadership could not solve them. I was seldom if ever looked to for solutions during garrison routine or on collective training. That is how good "a force generation Army" we have become. None of this adequately prepared us for what we faced in Afghanistan, where the essential moment of combat leadership occurred when soldiers could not conceive of solutions to problems during close combat themselves and turned to us for help. In those moments, we discovered the true value of being present – not watching on a screen from afar – to find the solution and demonstrate the strength of will needed to bring everyone to that solution collectively. It was at those moments when we discovered the value of years and years of training, professional development, education, and physical conditioning were important. It was there that I confirmed that not all leaders are equal, and that determination and intellect are the first requirements of command, not nice-to-haves.

In preparation for Afghanistan, I worked hard with the Regimental Sergeant Major Randy Northrup to get one thing right, that we select the right command teams from section to sub-unit level, on firm belief that personality and character of leadership was more important than anything else. This emphasis was affirmed in Afghanistan. After our first trials in combat, once we had settled down from first excitements, I trusted in the subordinate commanders because they had natural tenacity and robustness. I trusted that they would not succumb to their fears, but would combine their talent with the natural fighters under their command to enable the platoon or sub-unit to sustain and advance in the midst of chaos, violence, uncertainty and fear to fix and finish the enemy. I trusted that they would learn from mistakes and successes alike, that they would handle death and loss quickly, with dignity, and then move on. I trusted that they would enforce good discipline and order, and yet exercise forgiveness and compassion, without jeopardizing trust and lapsing into blame. All of these trusts combined produced the climate of mission command necessary to defeat counter-insurgency. None of it required sophisticated technologies. The emphasis the Canadian Army placed upon "big head, small-body", "intelligence-driven", "command-centric" operations disappeared in the lush green and highly compartmentalized terrain of Zharie/Panjwayi where good leadership based upon intellect, imagination, experience, determination and trust produced the effective destruction of the enemy. Above all, it was the constituent elements of trust that won. As we move forward in Army and CF developments, may we never lose sight of this again.

CHAPTER 12

IN THE BREACH: THE COMBAT COMMAND OF LIEUTENANT-COLONEL OMER LAVOIE

Colonel Bernd Horn

Most officers, and for that matter senior non-commissioned officers (Sr. NCOs) and soldiers, have spent the better parts of their careers preparing for war. They have read about the chaos, friction and sorrow of battle; studied tactics and operations; honed their technical warfighting skills; and participated in mock battle – all this to prepare them for the day when they would finally be tested in combat. For many, entire careers pass without that experience. For others, that quest is fulfilled. And when it is over, very few do not find it a life altering experience, for combat is the most dramatic of human endeavours. It brings out the best and the worst in individuals. It is filled with excitement and exhilaration, but also sorrow and loss. To those who hold responsibility for others – it is the ultimate test of confidence, loyalty and humanity. To them falls the task of mission success, which often competes directly with ensuring the well-being of others. In the end, they are the ones who have to send individuals into harm's way, time after time.

Lieutenant-Colonel Omer Lavoie is one such individual. He was the commanding officer (CO) of the 1st Battalion, The Royal Canadian Regiment (1 RCR) Battlegroup (BG) – designated Task Force (TF) 3-06, in Kandahar, Afghanistan from August 2006 to February 2007. Lavoie is one of the few combat experienced COs in the Canadian Forces (CF). He took over responsibility for the Canadian area of operations (AOR) in Sector South under contact in August and was engaged in constant combat for the next three months. Even after that, his tour was filled with continuous contact with the enemy. Within the first three months of operations in theatre, TF-06 suffered approximately 19 killed and 90 wounded.

This chapter is based on a number of discussions and interviews conducted with Lieutenant-Colonel Lavoie in Afghanistan, in October 2006. It contains the initial reflections of a battle-hardened CO. However, in the interest of promulgating lessons learned and capturing these initial thoughts within this volume, which was ready for publication, I have decided to use this format, rather than impose on Omer Lavoie who, at time of writing, was still heavily engaged in operations. As a result, any errors in interpretation rest strictly with the author.

As indicated earlier, Lieutenant-Colonel Lavoie conducted his relief in place with the previous battle group under contact and within weeks of arriving which, from 2-16 September 2006, was fighting a major conventional battle, which has come to be known as Operation (OP) Medusa. In its aftermath, TF 3-06 was heavily tasked with operations throughout Kandahar province, and Lavoie was consistently forward, sharing both risk and hardship with his dispersed forces. The lack of resources, harsh terrain and complex operating environment imposed an exhausting regimen on both the CO and his tactical headquarters. Often, mission success became a function of imposing personal will.

For Lieutenant-Colonel Lavoie, upon reflection of his first two months in theatre, there were a number of leadership challenges that he and his troops had to deal with. The first major challenge was that of unrealistic tasks and expectations that emanate from higher headquarters. In the western discourse of war, there are three fundamental catalysts that can affect the outcome or success of a mission irrespective of the reality on the ground. The first is a major scandal or series of scandals such as Abu Ghraib, and the rape/murders in Iraq, or in the Canadian case, the torture murder of a detainee in Somalia in 1993. A scandal of this nature can precipitously erode public confidence in its military and the respective mission in question. The second catalyst is unacceptable collateral damage to the civilian population in the conflict area and the third, and arguably the most explosive catalyst, is national (i.e. own forces) casualties.

Added to these factors is the impatience of the public. Current society has become more and more impatient and demanding of instant gratification. Spoiled by such rapid successes as the 1991 Gulf War and the 2001 invasion of Afghanistan, and increasingly ignorant of the complexities of war, the public often expects success and quantum leaps in results in weeks, if not days. Exacerbating this dilemma, are national governments that wish to give their temperamental voting public what they want. As a result, there is often pressure flowing from the highest levels downwards demanding "progress."

It is within such a context that Omer Lavoie was initiated into conflict. With political pressures at play domestically in Canada, as well as in the capitals of almost all contributing nations, there was an overwhelming desire by political and military senior leaders to show results, particularly in the realm of development, regardless of what challenges may have been faced on the ground. "NATO headquarters," Lavoie lamented, "often are unrealistic in their taskings – too often they have too great an appetite." He added, "I've had to say no on a number of occasions."

Anxious to demonstrate that progress is being made to placate demanding national governments and politicians, the pressure on field commanders can become enormous. "It's hard to convince headquarters of the reality on the ground," explained Lavoie, "They've never been forward – they have not visited the troops forward, yet they don't believe I can't start phase IV reconstruction even though I've lost seven guys on this road [Route Summit]."[1] Higher headquarters consistently tried to assign additional tasks despite Lavoie's forces being stretched thin to provide security over the road construction and other forward operating bases (FOBs) within his AOR – a direct result of a failure of host nation forces being made available to undertake static security tasks. Lavoie learned through hard experience that often the CO just has to say no. Furthermore, when promises of additional resources were given, he quickly discovered that it was prudent to wait until the promised troops or equipment arrived before beginning the task.

His dilemma at the time, i.e. his forces being tied up in static locations guarding the road and constantly under attack, was a direct result of taking on the task with the promise of Afghan National Army (ANA) forces being assigned to provide the necessary security. But, they never showed up. And when some did – their numbers were insufficient. "I've learned to say no to tasks until the manpower allotted to it shows up," acknowledged Lavoie, "Working/taking up commitments prior to the necessary resources arriving is why we are in our present mess – road construction and security."

Lavoie supports the idea that reconstruction is all important. "That will actually defeat vice just killing the enemy," asserts Lavoie. However, he was under constant pressure to get on with reconstruction because it was important for domestic, as well as international consumption. "But the fight isn't won yet," Lavoie explained, "the Taliban is still very strong and we must defeat them first for lasting reconstruction." As such, the CO, as well as other leaders on the ground, face an enormous challenge to provide security and create the conditions, if not perform the action themselves, for reconstruction.

---

[1] Reference to Route Summit, a new road being cut between Highway 1 and Ma'Sūm Ghar in Pashmul to increase security and provide a better road network to assist economic development. It was a direct result of operations conducted during Operation Medusa.

Part of the challenge was the limited resources available. National caveats of contributing NATO nations made finding troops to share the load in volatile areas such as Kandahar and Helmand provinces very difficult. In essence, only the Americans, British and Canadians could be counted on to conduct combat operations. However, they too are heavily engaged in their own volatile sectors. Given the huge AOR,[2] Lavoie found his Battle Group continually stretched thin. It was not uncommon for his troops to be in the field living in the dust, squalor and constant threat of attack for up to 24 days at a time. This stretched the nerves of the soldiers and did not allow for adequate hygiene. But, equally trying, it did not allow for adequate maintenance of vehicles or weapons. "The troops are constantly in the field," he stressed, "I'm waiting for ANA to assist so I can rotate troops in for 48 hrs R&R to decompress."

The resource problem stemmed from more than just NATO member nation intransigence to share the load and Host Nation difficulties in meeting troop assignments. The large number casualties of suffered during OP MEDUSA (e.g. 5 killed and approximately 40 wounded) cut a swath in the ranks as well. At the time, the reinforcement system was overwhelmed by the large number of casualties and it was taking up to 48 days to provide a replacement. In addition, the Home Leave Travel Allowance (HLTA) rotation system, which was based on sending complete sections at a time, also kicked in and further depleted the number of troops available.[3]

In the end, it was up to the CO and his leadership cadre at all levels to navigate the challenges in a high threat environment with the resources available. At times, it meant simply saying no. At other times, it meant creatively generating the necessary forces from the available pool. On more than one occasion, this meant that 9er Tac (i.e. the CO's four vehicle tactical headquarters) acted as fire support in a gunfight, vehicle recovery and escort, as well as casualty evacuation. It also meant that troops regularly had to dig deep and stay in the field just a little bit longer.[4]

---

[2] The AOR, Kandahar Province was 66,000 square (sq) kilometers (km). The BG focused on a 20 sq km area.

[3] The section leave plan was adopted from the previous tour where under the circumstances at the time it was found to be less disruptive and more efficient. The whole question of leave was a difficult one. No-one wanted to leave their comrades short in such dangerous conditions, but equally, the long demanding tour made the rest very important.

[4] The inability to rotate in to camp for rest and refit did not appear to be a huge issue. In fact, many troops took pride in the fact that "they were the longest in the field without a break." Concerns for vehicle and weapon maintenance in a very dusty, harsh environment, however, were always present.

The new security environment, specifically, the asymmetric threat and fighting a counter-insurgency was also seen as leadership challenge by Lieutenant-Colonel Lavoie. He explained:

> Probably the first and foremost challenge for a commander is actually getting your head wrapped around the idea that you're not in a linear battlefield. That drives everything – you are continually in contact. No matter where you are, you are always in contact. And even if you go back to what we consider to be our most secure area, which is the Kandahar airfield, guys there are still getting rocketed on a regular basis. It's a big paradigm shift – we're still a conventionally trained army but it's now changing with the new generation of the soldiers and junior officers that are coming in and deploying on operations. But certainly, you know, the training that most of us received has always been that you advanced towards the enemy and then everything behind you was the "rear" and the rear and flanks were generally secure. However, that concept is gone. We must change how we think about lines of communications, the rotation of deployed forces to a secure area where maintenance and rest can be done and the idea of securing bases to operate out of. A major focus now is just securing – securing lines of communications and securing forward bases. So it's just something that we're getting to know, but it's a huge challenge. The contemporary operating environment itself, specifically working within an insurgency, poses another whole range of challenges, such as my greatest frustration, the difficulty of gaining the necessary intelligence, as well as working in an environment where it is so difficult to differentiate friend from foe. For instance, I know right now that within this ANP [Afghan National Police] compound where we are working, there are Taliban operatives. And I've been in operations with the ANA, where I was later told by higher sources that that there were Taliban operatives sending messages back to the Taliban within the very ranks of the ANA officer leadership – officers that were part of my command post. So, you have to completely adjust everything you do and how you work. In addition, identifying insurgents in the field is no easier. One second you have a farmer in a field, the next he drops his shovel picks up a weapon and now he's an insurgent. So that whole environment is very difficult to operate in. And then, there is the question of asymmetric threats. We received excellent training, as much as you could expect, but still the idea that someone is willing to walk into a group of soldiers who are surrounded by a group of non-combatants and detonate a suicide bomb that

will kill soldiers as well as half a dozen or more civilians at the same time is still hard to accept. And that adds a whole new challenge. Soldiers have to be cautious and aware of the threat, but at the same time must try and win the hearts and minds of the population. There is nothing the Taliban would like us to do more than to shoot every civilian that comes within ten metres of us. If we did that – they win. I would say information operations are more important than the actual kinetic operations that we're conducting. That is information operations directed at the enemy, Afghan public, as well as our national public as well. The easiest method by which the enemy could defeat us is by convincing the people back home that we shouldn't be here and create national pressure to pull us out. Ironically, we understand the theory behind information operations, but the bottom line is we're still kinetic warriors.

Lavoie's assessment is telling. In counter-insurgency, body counts do not matter. As he has observed, "The Taliban can regenerate very quickly." He opined, "What counts is winning the centre of gravity, that is shifting the population to support the national Afghan government and the coalition." The crux of the problem, as Lavoie pointed out, is trying to win the hearts and minds of the people, yet, defend against the asymmetric threat. "We won't win with bullets," he counselled, "we'll have more success with other lines of operations" [e.g. reconstruction, development and governance].

The CO provided a graphic example of the new paradigm. "I'm building a road," he explained, "I never thought I would be doing this." Originally, his Battle Group were responsible for Phase IV Resettlement and Reconstruction. The ANA were to do security but they never showed up. Therefore, the only way Lavoie could provide a modicum of safety for his troops was to build a new road that avoided the only existing route, which was a narrow, constricted dirt road that snaked through close terrain. This existing path was a nightmare for the Canadian troops but ideal ground for ambushes, mines and improvised explosive devices (IEDs). In the end, however, the road is a major step forward. "In the short term," explains Lavoie it provides security for the region and the FOB." In the long term, he added, "it provides development and security, specifically, access to the area."

Gaining access is a huge issue in counter-insurgency, particularly access to the population. As such, the CO acknowledges that the Task Force is very sensitive of the culture. He also points out that the chasms in the very tribal-centric Afghan culture provides opportunities to leverage locals to get them on your side. "To a big extent," informed Lavoie, "when I was going into these villages trying to sort of put together who was who; who was with us and who was against us, the first thing I'd have my eye out trying to determine what was the different tribal make-ups in that particular group, as well as to determine which tribe it was that I was talking to." He added, "I would always talk to them separately as tribes, to get them to come onside and I'd try to use to tribal bonds and differences to the Coalition advantage because tribalism is pretty strong here."

In the end, Lavoie emphasized that a cultural understanding is key to counter-insurgency. "Certainly, in a counter-insurgency context, the understanding of the culture is critically important," insists Lavoie. "Often," he stresses, "You can turn the population from being black or anti-coalition to pro-coalition by simply understanding the cultural sensitivities and by massaging those in your favour." Conversely, he explains, "you can very quickly turn them the other way by inadvertently not being sensitive to a lot of their cultural sensitivities and inadvertently reinforcing the negative stereotypes or Taliban propaganda."

The subject of Afghan national security forces (ANSF), is another leadership challenge and critical issue to understanding the nature of counter-insurgency operations in Afghanistan. The CO believes a major component of his mission is to build the security capacity in the ANSF (i.e. ANA and ANP). The ANA is a relative success story despite ongoing challenges. Overall, the will to fight the Taliban is omni-present. Despite their lack of equipment, poor training and pitiful pay, they are brave and courageous warriors, albeit ill-disciplined soldiers. The greatest problem with the ANA is their lack of numbers. Moreover, there is still some work to be done in regards to coordination and cooperation. For example, there still exists some confusion over terminology. ANA forces may be under "operational control," however, once the ANA are given orders, they must still clear their task with their Brigade commander. For example, the CO directed ANA troops to dig-in on the road to provide security. Once so instructed, the on-site subordinate ANA commander relayed the order to his higher headquarters for approval. This created a delay of one and a half days. Nonetheless, as Lavoie quipped, "compared to the ANP, these guys [ANA] are Spetznaz." He explained, ANA at least wear a uniform and usually they don't shoot at you."

The ANP are currently a major weak link in the counter-insurgency effort. They are less numerous and less effective. Most receive two weeks of training and are extremely young and inexperienced. Lavoie quickly discovered that the ANP battalion commander with whom he was to work was illiterate and unable to read a map. "I had to take him out to tell and show him where to place his troops and tactical control points," stated Lavoie. In addition, he added, "the ANP don't wear uniforms because they are 'uncomfortable' so there are ANP in civvies carrying AK-47 assault rifles in unmarked civilian pattern pick-up trucks." An exasperated Lavoie stated, "Its hard to tell good guys from bad. For that reason there were some fratricides early on in the operation."

The ANP is intended to be an integral component of the counter-insurgency campaign. They should represent the national government and help provide security. However, they fall far short. Their checkpoints in the TF 3-06 AOR are designed to support security through observation and physical security. Yet, the CO and others have walked in on checkpoints where all members have been sleeping with weapons totally accessible or where many are often under the influence of drugs. In addition, their poor pay and lack of professionalism make them open to corruption and crime, which further alienates them, the national government and the coalition, with which they are seen aligned, from the people.

The next major leadership challenge is the reality of combat. "I would challenge anyone," declared Lavoie, "to tell me that I'm not at war." Few who have spent time with him or his troops would disagree. "The biggest shock," reveals Lavoie, "is that you will lose guys and you are losing guys." He elaborated:

> It's a bit of a shock and as a leader you must put it into perspective. But, it's a huge morale issue. Soldiers see their friends, buddies and colleagues on the ground dead. They don't see the 200 enemy dead. The guys are definitely hurting. I'm just not sure how you could ever replicate that in training. I mean we certainly covered the issue in our preparatory training back in Wainwright and other places. You can get your TTPs [tactics, techniques and procedures] down for doing the casualties, but you never get that true battlefield inoculation of actually seeing 37 soldiers laying on the ground all wounded, all in pain, all in agony, all needing treatment, as well as the dead along with that. And that's just something that never gets any easier. I think your TTPs – the procedural stuff, can be ironed out and more

efficient SOPs developed, but the psychological side of it never, never gets any easier that's for sure.

Operating within a coalition provided other sources of challenge. Firstly, there is the issue of national caveats as already mentioned, which limits who is willing to send forces where within the country. This often results in a few nations doing all the heavy lifting (i.e. combat operations). Secondly, as Lavoie states:

> They're all the challenges of working within any coalition [e.g. national agendas, different SOPs, TTPs, languages, staff procedures], however, the biggest challenge in Afghanistan is that you very quickly learn that NATO in itself has virtually none or very little of the necessary combat enablers. For instance, I briefed the commander of ISAF personally on the context of my operations for OP MEDUSA. He in turn confirmed to me that I was ISAF's, no actually he said NATO's, main effort. As a result, one would think okay, it seems like the commander supports everything so I should be given the support I need. However, that isn't necessarily the case because NATO doesn't own the enablers. The enablers for the most part are still owned by their contributing countries and here in Afghanistan, largely the enablers that we're looking at are aviation, air and ISR [Intelligence, Surveillance and Reconnaissance] and they are still American or British, and to some degree in this theatre Dutch. So, even though he says you are the main effort, you still have to convince those countries that you are the main effort. And if they are reluctant, NATO doesn't even necessarily have the hammer to direct, for sake of example, the British to provide this and the Americans to provide that. So that was the first real big thing I noticed. As a result, there was a lot of begging. I know that the Multi-National Brigade and even the Commander of ISAF had to go and do a lot of groveling to and bartering to get the required assets shifted over to our operation. And the effect it had on me though was that the enablers that did become available had restrictions on them... For instance I was told, okay you can have these things for X-amount of days, but then they're being shifted back to whatever region again.

Another huge challenge and frustration Lavoie dealt with was dealing with pressures from commanders from other countries to do things that were not in consonance with Canada's best interests, for example operations that may result in higher than acceptable friendly casualties. "I've had to say no," revealed Lavoie, "I'm not doing that because I didn't think it was in Canada's strategic interests particularly in light of Canada's criteria for mission failure. He added, "there was clearly things that they wanted me to do that within my assessment in terms of friendly casualties forced me to literally say no, I don't think that's a good idea."

Part of the problem is the divergent views on acceptable casualties. For some countries it is higher than others. "There are always members of the coalition whose acceptance levels are greater than or, or less than ours," Lavoie asserted, "I have certainly had pressures from above from multinational countries whose acceptance of casualties was a lot greater than Canada's and I didn't accept the task – I wasn't prepared to take that degree of risk or that many casualties." Conversely, he added, "At the same time I had sub-units attached to me whose acceptance for casualties and risk was a lot lower than ours." He explained, "So to get those attached sub-units from other countries to do what I wanted them to do was really difficult."

Although counter-insurgency is a major focus of Lavoie's mission, he was quick to point out that the CF, or any military for that matter, cannot lose its conventional warfighting ability. "We're here to conduct counter-insurgency operations and having said all that, this tour at least," he commented, "has shown that we obviously should never ever lose the ability to conduct the full spectrum of conventional operations that we've always been grounded in, up to and including battle group operations."

Lavoie's observation was rooted in his experience. His first major operation, for all intents and purposes, was conventional battle. OP MEDUSA was a major offensive conducted from 1-17 September 2006, by ISAF with assistance from the ANA. Their objective was to establish government control over an area of Kandahar Province centered on the town of Panjwayi, approximately 30 kilometres from Kandahar City, which was the birthplace and heartland of the Taliban. In essence, it was an enemy stronghold. NATO's intent was to separate the tier two Taliban (local hires) from the tier one (hardcore) elements and either destroy or capture them. This was designed through a phased approach engaging the local leaders first and then applying superior and precise combat power. By 2 September 2006, it became apparent that fighting would be inevitable.

When the operation commenced, it seemed as if it were a Cold War training exercise. "There was nothing new in warfighting," quipped Lavoie, "We used air to hit deep and danger close artillery at 300 yards or less." He added:

> Dismounted infantry rushed in before the smoke cleared and seized the objective. Engineers cleared a route with bulldozers and dealt with IEDs. We pushed LAVs [Light Armoured Vehicle III] up to support infantry to the next objective and to the next bound. At night we conducted fighting patrols and Reconnaissance Platoon seized the line of departure for the next bound the next day. And much like predecessors in Vietnam who said they had to destroy the village to save it – we had to do the same. Welcome to my world.

However, the difficulty of combat operations quickly became apparent. Pashmul was a green belt with thick vegetation. This, combined with its close proximity to the Pakistan border, provided ample cover for the Taliban. The Taliban chose their ground well. Seven-foot high marijuana fields hid movement and masked the thermal imagery of the LAV. Within Pasmul, they developed a sophisticated strongpoint replete with entrenchments that resembled a Soviet defensive position. Communications trenches were dug to connect the larger trench system and bunkers. Moreover, the existing canal had been widened with light equipment to act as a tank trap.

Lavoie took the high ground first, namely the Ma'Sūm Ghar feature, from which he could dominate the area by both observation and fire. His battle group was originally given eight days to clear the objective. The Taliban positions were pounded by air and artillery and good results were attained during the first two days. An assessment by higher headquarters now sped up the process. "They wanted to rush it through," stated Lavoie, "so I asked what happened to the original timelines." He added:

> Consistently higher pushes. They want IO successes, and brigade in turn is pushed by NATO HQ for obvious success, however, the calls for greater effort and faster results never come with resources. Since there are not sufficient resources, you must phase the operation, therefore it takes longer. Now, I just say no – I learned my lesson the hard way at the beginning of the operation.

In the end, the attack was successful. "The Taliban didn't think we would attack from the North," explained Lavoie, "because the ground was impassable and no-one had done that before." A clever deception in the south and the use of bulldozers to plough lanes through grape vineyards and Marijuana fields to make lanes for the LAV IIIs, as well as "B" Company utilizing the ground in conjunction with artillery bombardments at "danger close" range, overcame any and all opposition. By 17 September 2006, OP MEDUSA was successfully terminated.

OP MEDUSA was NATO's, as well as Lieutenant-Colonel Lavoie's first combat actions. Retrospectively, he assessed:

> Well, I think first and foremost is the requirement to ensure at all levels that the battle, particularly for us the Canadian component, is fully thought through to the end-state. And, that means it must be very clear. For instance, we may have to embrace the idea of unlimited exploitation. Lines on the map look good – based on our conventional training in this case the lines looked about the right sort of amount of area that battle groups would take on. But they didn't account for the next steps forward, the terrain or type of conflict we're fighting. What we accepted as exploitation, what we all accepted and recognized as just a line on a map did not correspond to reality on the ground. The insurgents don't respect that line. So in thinking through the whole battle and, and looking beyond my own battle, which really just amounts to be my areas of influence and interests, in hindsight, we should have looked beyond the doctrinal approach to exploitation. Now we may not have been able to do anything about it because the engaged units and formation only had a set number of troops and resources, but perhaps we may have been able to change the plan to the point at least where we don't stop at Phase IV, but just keep going and going until the whole campaign is complete. For example, in my case I think we should have exploited another 20 km to actually seal off an area to disrupt their [Taliban] lines of communication. There would have been a lot more fighting at the time, but we had them on the run and we probably could have kept that pursuit and then actually held the ground we seized and put the Taliban in a position where they would not have been unable to come back into the area. That would have saved us from our current position where we are not going back into the same ground we had already cleared. As a result of this thinking I find

myself in a position where I am challenging the instinct of the higher headquarters. I'm trying to impress upon them that we just don't want to disrupt Taliban operations, we actually want to clear and secure the area and prevent the, the reoccupation down the road. In the end, its still linked to our conventional method of thinking – take the enemy position and then just move it on one step forward. In a non-linear battlefield, exploitation is much more important and more complex. It's not about just capturing a piece of ground.

On the positive side, the CO assessed that he learned some things very quickly. For instance, he commented that he "learned three hours into the transfer command that although the insurgents have a lot of advantages over coalition forces in terms of being able to blend into the countryside and their knowledge of local terrain, population and culture," he concluded that "we have a whole lot of strengths that we can pit against their weaknesses. As such, during OP MEDUSA, he utilized this realization when he crafted his commander's intent paragraph for the operation order. Specifically, he stressed the requirement to disrupt the Taliban decision cycle based on the fact that Coalition forces had the ability to operate across the entire battlespace, as opposed to the enemy, who was restricted to working within the confines of specific battlefields. For example, Coalition forces operated in the air, across the electronic spectrum, as well as everywhere on the ground. Conversely, the Taliban was restricted to operating only along the ground, and even there, with limitations at times.

Lavoie also examined the initial contacts. He noted that the Taliban had very poor kinetic control, i.e. the ability to deliver consistently – timely, accurate and responsive fires to bear on the tactical situation. Conversely, as he fought the battle, he was delivering huge amounts of kinetic effects through his ground forces, through indirect fire and close air support. In addition, he was exploiting the electronic spectrum. Through his ISTAR capability, he could often adjust to Taliban manoeuvres and react in such a way as to disrupt their decision cycle. For instance, through Unmanned Aerial Vehicles he could watch them manoeuvre to form an ambush, "which I could then counteract by firing hellfire missiles into that ambush site, followed by artillery." Furthermore, through their situational dominance, they could detect the enemy massing on mountaintops, or moving to a specific road junction, and before the Taliban even reached their destination, they would be bombarded with artillery. As a result, asserted the CO, "we're able to keep just ahead of the their decision cycle and prevent them from getting within ours and as a result we can normally outmanoeuvre them." He added, "Even though they had a 6-1 manpower ratio over us they could never bring their superiority in numbers to bear or concentrate their force."

Another lesson learned Lieutenant-Colonel Lavoie hoisted aboard was in regards to airpower. He explained:

> I sort of kick myself for having to relearn the lesson in regard to airpower because it's something I think every army officer, army soldier knows intuitively, that is, that you cannot win with air power alone. And in the Battle of Pashmul, I just got sucked into it – I probably had an exorbitant dependence on, or overestimation of air power. We dropped thousands of tons of munitions on them in the three or four days before we had to start conducting the ground operations. But in the end, as we relearned, you just cannot rely on it because until you actually get in there and muck out those fortified defensive positions you just can't be sure. It's a matter of getting those ground forces in there and take them out. It was hard for us not to get sucked into it because you were sitting on the high ground watching our forces just closing them down with 25 millimetre cannon fire and the soldiers were seeing just literally hundreds of thousands of tons of bombs being dropped on the objective. You couldn't just help thinking that there was nothing that could survive in there. Well we found out the hard way that they could. That is a lesson that we probably relearn in every modern battle that we have fought since we started using air power. So, you just can't overestimate the effects of air power on a low tech enemy.

Lavoie further philosophized that hand in hand with the requirement not to overestimate the effectiveness of airpower was the requirement not to underestimate the tenacity of insurgents, "especially a radical fundamentalist idealist of an enemy because." He recalled:

> You would just sit up there and you'd be taking your turn in the turret with the 25 millimetre cannon. In the first 48 hours of the engagement you never had more than one or two of my LAVs at a span that were not engaged with targets. And that was because the Taliban just kept on coming. And you'd just think at some point they're going to stop – they're going to get it and not come anymore, but they just kept on coming and coming and we kept engaging them. At one point the troops called it a shooting gallery.

The last point Lavoie commented on was the ability to "cut-off" the enemy. He noted that in an insurgency this is very difficult to accomplish. "That's probably one of the weaknesses in the plan overall," he conceded, "there was never really a plan to cut them off, to prevent them from seeping out and escaping." The Coalition forces pushed from the north and the south east, and tactical aviation was used to tighten the cordon. However, Lavoie points out that there was a lot of reliance on tactical aviation to prevent the Taliban from escaping to the west. "Although tac aviation intercepted, killed and destroyed quite a few of them," acknowledged Lavoie, "there were still quite a lot who escaped." He added, "we really needed a ground force to be put in there, but it comes down to the fact that we really didn't have the forces to do it."

The other problem he noted was that of identifying the enemy. "I had a sub-unit screening my left flank 15-20 kilometres away and their job was to contain the seepage," he stated, "but they could just report that 30 fighting age farmers were leaving the area, but in accordance with their rules of engagement, because they were unarmed, they couldn't engage them." The frustrated CO questioned, "what could you do to these guys." He added, "You were always certain that they were Taliban, but they're just walking down the highway."

Importantly, Lavoie noted that it is cases such as this where host nation forces from the ANSF are critical. They are cultural familiar with their fellow countrymen and always seem to have a sixth sense. "These guys [ANSF] either through intuition or certainly after talking to suspects, just know that they're not farmers," explained Lavoie, "we [Coalition forces] just can't do that." Its for that reason, argues Lavoie, that all operations must be done in conjunction with host nation forces. "We need the ANSF to act as cut-off groups," he insists, "so that they can question and identify individuals fleeing a combat area."

This short chapter is not meant to be a definitive study on command in combat. Rather it is an initial reflection on some of the leadership challenges faced by a commanding officer in the new complex security environment. The impressions and reflections shared, are done so to provide others with a degree of vicarious experience and an opportunity to reflect on the challenges they may face in similar situations. In the end, it is incumbent on all military commanders to seek out as much current and relevant knowledge, experience and lessons as possible, so that they are better able to lead Canada's sons and daughters into harm's way.

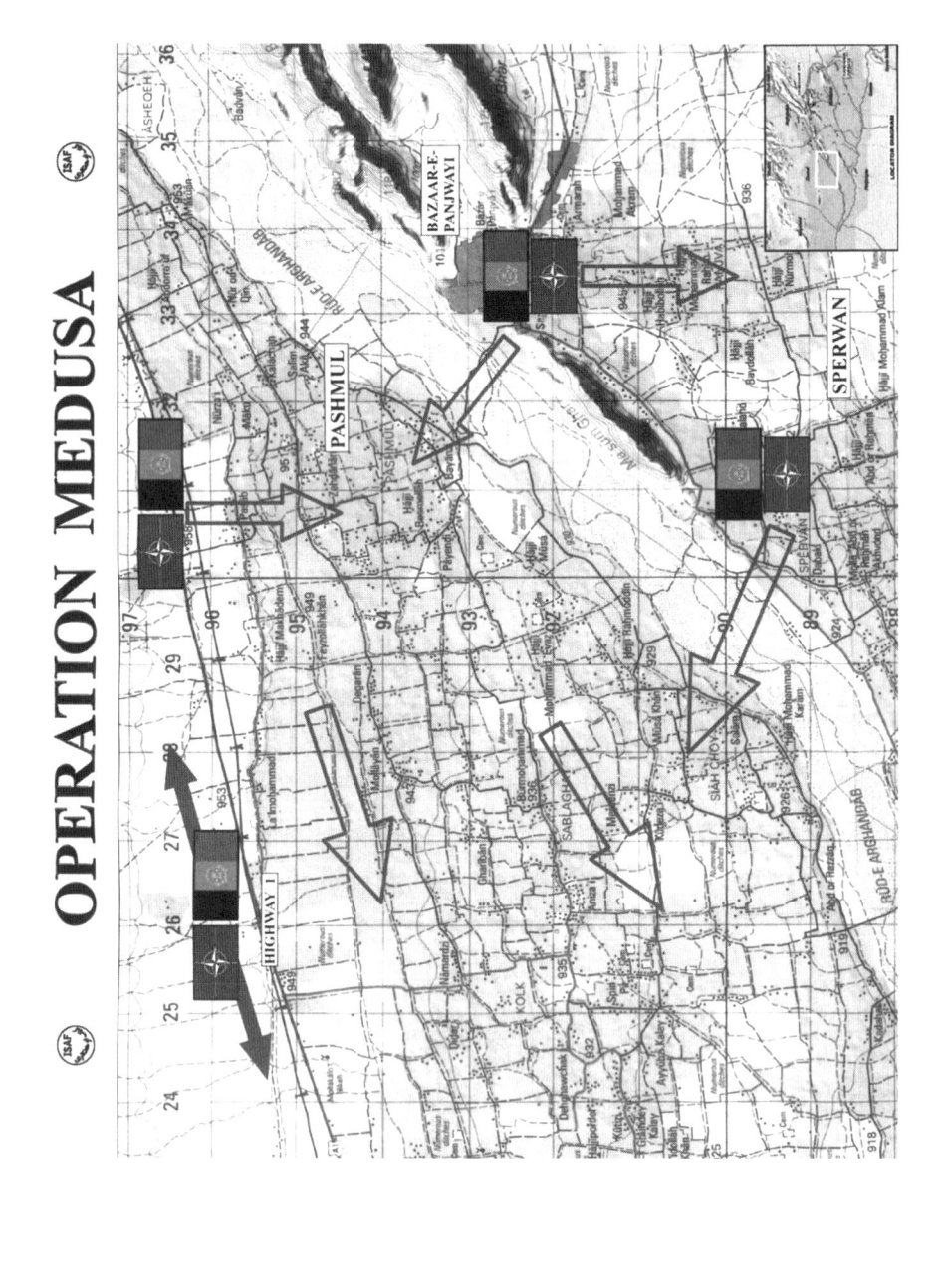

OPERATION MEDUSA

CHAPTER 13

NO SMALL VICTORY: INSIGHTS OF THE COMMANDER OF COMBINED TASK FORCE AEGIS ON OPERATION MEDUSA

Brigadier-General David Fraser

*****

Editor's Note: This chapter is based on an interview conducted with Brigadier-General Fraser on 21 October 2006, at the Multi-National Brigade (MNB) headquarters (HQ) at the Kandahar Airfield (KAF). His narrative is an introspective look at NATO's biggest (albeit only) battle to that point since its inception. Operation Medusa was a major offensive conducted 1-17 September 2006 by NATO's International Security and Assistance Force (ISAF) with assistance from the Afghan National Army. Their objective was to establish government control over an area of Kandahar Province centered on the town of Panjwayi, approximately 30 kilometres from Kandahar City. This area was the birthplace and heartland of the Taliban and was considered an enemy stronghold. Nonetheless, NATO forces intent was to separate the tier two Taliban (local hires) from the tier one (hardcore). This was designed through a phased approach engaging the local leaders first and then applying superior and precise combat power. The most dangerous course of action, which was exercised by the Taliban, was to stand and fight on ground of their choosing and advantageous to the defender. In fact, they built bunkers and tunnels and carefully prepared kill zones to meet the coalition offensive. Combat action started on 1 September with air strikes and artillery bombardments that lasted approximately 48 hours. On the third day, an attempt to advance into the stronghold area was met by devastating fire and NATO troops pulled back. The following morning a catastrophic friendly fire incident in which an American A-10 Thunderbolt aircraft accidentally strafed Canadian troops leading to massive casualties halted combat operations for another day. On the following, two days air strikes and bombardments again pounded the Taliban positions. There was then a brief lull in the fighting, with renewed combat from 8-10 September. In the following days, ISAF forces began to redeploy and squeeze the Taliban throughout the area. After suffering huge losses (estimates run at 1000 dead), the Taliban quietly withdrew under the constant advance of the NATO troops. On 14 September ISAF troops began to push into the last of the formerly controlled Taliban areas.

Three days later, NATO announced the operation completed. NATO spokesman declared Operation Medusa a great success in destroying Taliban forces and heading off a planned enemy offensive designed to capture Kandahar City. In the end, Operation Medusa had more in common with conventional warfare than it did with counter-insurgency operations. Combat operations were to be complimented with reconstruction of infrastructure and economic assistance to the local population. This was done; however, the main effort shifted to reconstruction following combat operations. Taliban forces quickly began to infiltrate back into the area and they adopted an asymmetric approach using suicide bombers, improvised explosive devices and ambush tactics. As a result, reconstruction in the immediate aftermath of Operation Medusa was slow and combat actions continued. As the commander of the MNB/Combined Task Force (CTF) Aegis, Brigadier-General Fraser was responsible for the inception, planning and execution of the operation. Here is his story.

*****

We had been in Panjwayi a couple of times before with Task Force (TF) Orion and we knew it was an important area for the Taliban. However, we didn't have enough forces to go in there earlier for a sustained offensive effort, other than what we conducted with Lieutenant-Colonel (LCol) Ian Hope's battle group. But the scale of Taliban activity made it clear we would have to come back. In fact, it was 3 August when we went in there and lost a couple of guys. We received intelligence of major activity, that's why Hope was going back in there. We knew it was big and LCol Hope went in and got hit hard in the objective area. We met after that and analysed what happened. The Taliban used two improvised explosive device (IED) attacks roughly in the centre of the objective area, as well as RPGs (rocket propelled grenades). Our assessment was that the Taliban had gone into the Pashmul area, specifically the Bazaar-E-Panjwayi area, dug in, and now he was prepared to fight. Quite frankly, the Taliban was everything we thought and more. In fact, he was more sophisticated at what he was doing than we originally thought. So we analysed all that, figured out where he was and what he was doing and came to the assessment that he was acting as a conventional force. We then asked ourselves one question: what was the enemy's intent?

The enemy's intent, from our assessment, was to isolate Kandahar City, not directly but indirectly, to demonstrate the weakness and the inability of the national Government to come after them with a conventional force.

This also indicated to us that the Taliban were actually progressing with the evolution of their own operations to the next stage* where they thought they were capable enough to go and challenge the national government and coalition forces in a conventional manner.

We also assessed that their intent was to engage the international community in a battle of attrition on ground of their tactical choosing to cause as many casualties as they could to attack our centre of gravity (i.e. domestic public support). All of this was designed to defeat us from a 'political will' point of view; to illustrate weakness in the Government and thereby set the stage where the Taliban could attack the city and defeat not only the provincial government there but also attack the national government in Afghanistan in a fairly sophisticated and substantive way.

The other thing I noticed with the operations in August was that this felt familiar. I had read *The Other Side of the Mountain* by Lester Grau on the Soviet experience in Afghanistan during the 1980s, so I grabbed the book (it was in my room) opened it up and low and behold, there was Kandahar and the Pashmul area mentioned, not once but several times. The Soviets fought in this area and the more we looked into the history, the more we found that the ground may not be tactically important to win, but for years or tens of years, the area has been critically important operationally and strategically to the Mujahedin and Taliban.

So I made an assessment and I thought, "okay, they've gone conventional, this is their intent, so how do we defeat their intent." Well, I decided that we will defeat their intent by putting our forces all around them and we will wait them out. You see, they wanted us to get into a battle of attrition, to slug it out, to try and clear them out of that complex terrain where they have all the advantageous of a well dug-in and protected force – where our technological superiority could be nullified. I directed that we would wait them out. I reversed the roles on them. The Taliban went conventional and ISAF went unconventional. I decided that we would manoeuvre, feint and

---

* This refers to the Maoist model of insurgency: Phase 1 – Strategic Defence: focus on survival and building support. Bases are established, local leaders are recruited, cellular networks and parallel governments created; Phase 2 - Strategic Stalemate: guerilla warfare ensues. Insurgents focus on separating population from government; Phase 3 – Strategic Offensive: Insurgents feel they have superior strength and move to conventional operations to destroy government capability.

slap a cordon around them. We would engage them in a battle of attrition, but it would be on our terms, namely a battle of attrition through joint fires.

To achieve this, I had to constitute the appropriate force because I was short of soldiers. As a result, the British and the Dutch sent troops that were able to take over certain outposts and garrisons, which in turn freed-up the Canadian troops from TF 3-06 (i.e. 1 RCR Battle Group) so that they could concentrate themselves in Pashmul to conduct the actual offensive. In addition, I asked for Task Force 31 (American Special Forces) if they could go and conduct operations to our south-west near Sperwan. That way I could concentrate my forces on the main effort in the Pashmul area.

We anticipated that the enemy had two courses of action. One was that they would just continue to move around and we would continue to attack them. The second enemy course of action, the most dangerous, was if they attacked. This is what they did – they continually challenged us on the fringes of the terrain that held and fortified. Nonetheless, I wanted to wait for two to three weeks, all the while hammering them with fires and then eventually when I thought the time was right, when the enemy was physically and psychologically weak, then go in and seize the objective areas we had identified in the Pashmul area.

Therefore, when we went back into the Pashmul area, we had essentially two and a half battle groups out there – Task Force 31, which was a special forces battle group working to the south-west in the Sperwan Gar area to prevent enemy reinforcements from moving north, as well as increasing pressure on the Taliban; the Canadian TF 3-06, based on the 1 RCR Battle Group, focusing on the Panjwayi area; and two Kandak Afghan National Army (ANA) units who assisted operations. I also later created Task Force Grizzly to operate in the south as a feint.

The Dutch took over some outposts on Highway 1, which freed up additional Canadian resources. In essence, I had distributed my forces around this area to provide as much containment as I could. We adjusted that containment force around to get the enemy to move inside that circle so we could shape the battle and advantageously engage the enemy. We also dropped psychological posters to say look, [non-combatants/civilians] get out of there – our fight is with the Taliban not with you. Most of the people had all read that and left prior to the battle.

Let me just back up one step, we also had a key leadership engagement as part of the campaign planning. Our plan called for a four-phase operation. Phase one was the intelligence preparation of the battlefield. Phase 2 was shaping the battlefield, which included key leadership engagements, the actions I spoke about earlier. Phase 3 was the actual operations to clear out the Pashmul/Panjwayi area where the Taliban had fortified and dug-in and Phase 4 was for reconstruction.

In Phase 1, we spent a lot of time planning and gathering intelligence. A lot of effort was devoted in Phase 2 to building up – assembling the enablers and forces we required, as well as the logistical support. In addition, we attempted to lure the Taliban out so we could determine their exact size, location and engage them. Finally, Phase 3 was going to go when we decided we were ready and the Taliban were severely weakened.

Initially, we thought there were 500 enemy in the "pocket," 200 hard core, 300 tier two [i.e. local hired guns versus ideological fanatics] and my intent of Phase 2 was to engage the Taliban forces over a prolonged period of time with lethal fires (e.g. fast air, artillery). That would entice the less dedicated, you know, the tier two types, to give up the fight. I wanted to impact those individuals who had joined up to go in for a couple of days, you know, get paid a few bucks, have a few wins [successful attacks/inflict casualties on coalition forces] and then leave. I was going to draw out the fight with long range fires as long as I could and go after their minds, realizing that it would then be harder for their commanders to keep them motivated and keep them going. So this was all about time, patience, perseverance and not rushing into it. We had no intention of rushing into it.

In fact, our intention was to separate the Taliban by putting the pressure on them. We had them contained, they were fixing themselves – they had fixed themselves. They were bringing forces in from everywhere – infiltrating through the Red Desert, up from Pakistan; they assembled a lot of commanders in the pocket and we were controlling the agenda. The only thing that the Taliban had to decide was if they wanted to speed up the agenda, which they tried to do at the end.

So Omer [LCol Lavoie, CO TF 3-06] had a few bad days out there. The worst was on 4 September when he was preparing to attack in the morning, and about 20 minutes before he was to launch we had a friendly fire incident. An A-10 Thunderbolt ground support aircraft became disorientated and accidentally strafed "Charles" Company of the 1 RCR BG. That hit us hard with one killed and over 30 wounded. Basically, one of our sub-units became ineffective.

As a result, I created Task Force Grizzly. I reformed the remainder of "Charles Company, and the Americans sent their national command element and a rifle company. Based on the results of our first probe [on 3 September], it became clear that I needed to adjust because now, the enemy had focused on where he thought we were going to cross the Arghandab River from the south. I received an updated intelligence assessment from higher and it coincided exactly with our assessment. In short, the assessment was that this was key terrain tactically for the Taliban and they had reinforced and defended the northern shore of the Arghandab River.

In fact, they actually created a kill zone in the objective area we had code-named Rugby. They had designed to take us either from the east or from the south-east. They were really focused on us coming from the south and a lot of their commanders were in Kandahar, Sperwan Gar and along the southern part of the Arghandab River. Sperwan Gar was an important area because from there they could escape into Helmand Province or into the Red Desert, which led to Pakistan. As a result, based on that assessment and the initial moves by TF 3-06, I decided to give Task Force 31 the task to take Sperwan Gar and once that was accomplished push across the Arghandab River and take Siah Choy.

I created Task Force Grizzly to create the impression of force, but more importantly, it was actually a feint – to deceive the Taliban that we were still coming across the Arghandab River from the south. Meanwhile, I shifted TF 3-06 to the north so that they could initiate a deliberate sweep from our Patrol Base in the north, clearing our main axis of advance down to the Taliban strongpoint.

So, once that redeployment was done, I ordered TF 3-06 to begin putting pressure on the enemy and to do it in a very deliberate way. As a result, what the enemy saw, was more pressure being placed on him instead of us (i.e., we were not moving into ground of his choosing – we were forcing him to react to our choice of terrain). At the same time, I ordered Task Force 31 to move north into Sperwan Gar at the same time as the 1 RCR BG started moving from the north down – clearing towards the

Arghandab River. They were clearing the ground step by step. A Dutch sub-unit (brought in from the west of the province) was taking care of a patrol base in the north and Highway 1 and we had special forces down in the Red Desert. We maintained our cordon because what I was looking for was those 300 tier two fighters to leave, and the 200 hardcore fanatics to stay and continue to fight the battle of attrition. But, because I did not want to meet their intent, I decided to use manoeuvre and surprise.

The dates are a bit of a blur [see editor's note at start of chapter], but we got to a stage where in fact I would say there was a tremendous amount of pressure from ISAF to "get it done!" On the other hand, there was a tremendous amount of support from Canada to do the right thing but not to rush into it. And then, I knew there was a culminating point – how long could I keep my forces out here, and of course, the enemy had a major vote in all of this.

As the operation unfolded, Task Force 31 had a hell of a fight that lasted over three days. Task Force Grizzly was doing a great job keeping the enemy preoccupied in the south while TF 3-06 just cleared down from the north. As we continued the attack, we started to see the enemy, over time, he actually conducted at least one if not two RIPs (relief in place) and the intelligence was telling us, despite the attack and heavy bombardments, we were actually seeing fighters who were risking staying in place. The command and control was still very effective and still pressing very hard for the fighters to keep on going, even though they were taking a pounding.

Despite severe opposition, Task Force 31 was having success in Sperwan Gar. At the same time, Task Force Grizzly was doing what I wanted it to do and they were meeting with success as well. Then we made the decision to press really hard now. I mean, it's a feeling in a battle – you can feel the battle when you got the enemy. Its something you cannot teach, you just got to know when to push. Our forces got to that stage where you read the intelligence, you read what the soldiers were doing on the field, and then you just realize it – okay, it's time to push. And we went out there and we pushed because the enemy was starting to pull back, I mean we were not in any great strength there, TF 3-06 were making headway to our interim objectives and Task Force 31 were pushing as well, even though the enemy was just dumping on them at Siah Choy. In fact, the Taliban were coming across the intervening ground in convoys of trucks, dumping off 5 to 10 guys from each truck, all who just unloaded and attacked Task Force 31's position. It became a turkey shoot. In one night, I think they killed between 100 to 200 Taliban, it was a phenomenal shoot.

So I said, okay, we got something really big in Siah Choy and in Sperwan Gar, and there is little pressure coming out of the main stronghold so I pushed TF 3-06 to get down and take our main objective, which we called Rugby, in Panjwayi. At the same time I told Task Force Grizzly to get across the Arghandab River get into eastern side of Rugby – I told them to get across and then roll up the Taliban position from the flank, realizing that the Taliban would then collapse.

The plan did not unravel quite so easily. I gave the instructions to Task Force Grizzly to get across but they ran into some problems. They eventually got across with some national police elements, so I pressed the TF commander to get "Charles" Company minus, supported by elements of ISTAR [intelligence, surveillance, target acquisition, reconnaissance] squadron, across. They managed to do so against some medium resistance. Task Force Grizzly got into the enemy trench line and that's when I told TF 3-06 to push hard and link-up with Task Force Grizzly because once we had the momentum going and the Taliban started to fall back, I just wanted to keep the pressure on.

At that stage of the game, we had great pressure on the enemy – we were coming from the north, from the south and from the south-east. We had three task forces that moved in with significant pressure. Lieutenant-Colonel Lavoie and his 1RCR BG linked up with Task Force Grizzly, which was pushing towards them from the south-east to the north-west. Meanwhile, the moment Sperwan Gar was taken by Task Force 31, I ordered them to push to Siah Choy. We thought we were going to have a huge fight in Siah Choy based on our experience and the Taliban tenacity at Sperwan Gar, which was just staggering. As a result, I told the other two Task Forces to just stand by because the main effort that morning was going to be Task Force 31 and their push to Siah Choy. I allotted them priority on artillery, aviation and everything else.

I received word that morning that we took the town without a shot. Nobody was there, therefore, at that point we moved into exploitation. Amazingly, its hard to exploit, its hard to get troops to take more risk. Once you get soldiers going at a certain speed, to get them to change that speed and exploit is difficult. You read about it in the books and you think, how hard can that be – but it is. I went out and I talked to the commander of TF 3-06 and said now is the time to get all your forces and exploit – that means you take more risks but you don't take more chances or become reckless about the risks you accept. But, now you have the enemy on the run and now is the time to take the ground. And, I wanted TF 3-06 to take that ground because of what they lost there – they took some heavy casualties and I mean for them, there was psychological value in that terrain. In fact, I was there when "Charles" Company actually took it and I was so happy that they were the ones who went across and seized the ground – I thought, "you took it, no one else did, you guys did it" and that was another reason why I assigned them to Task Force Grizzly in the south – this was important for that company because it came at a high cost and they were all trained to take that ground. Essentially, the enemy, after all that pressure, after all that time, the enemy just collapsed and they went to ground.

Now, from all of that, I mean, how many people did we kill? I don't know. There are numbers out there and they all read well for history novels. We think we probably killed about 300 to 400, captured 136 (detained by the Afghan Security Forces) and we probably took out 5 senior commanders on the ground. A significant defeat, the worst defeat the Taliban ever experienced in probably 40 years according to the Afghan Minister of Defence.

The ISAF commander was ecstatic. He just could not believe what we were able to accomplish. He was very enthusiastic, I mean psychologically, what our troops did was impressive. They saved the city of Kandahar, arguably saved the country and they saved the alliance. They proved that NATO could fight as a coalition.

The staffers at brigade headquarters did marvellous stuff. I'd do it again the same way: smart, slow, deliberate. Slow not because we were slow; slow in that, that was my intent. I did not want to do battle with the enemy on his terms. Rather, I planned to counter his intent. I slowed the tempo down to allow us to use long ranges fires to attrite him to avoid rushing in to attack an enemy who was waiting and prepared in fortified positions. There was no rush. The enemy was fixed. In fact, we welcomed his reinforcements because we knew the in and out routes and by reinforcing the pocket the Taliban set themselves up for a bigger defeat.

That's all the good stuff. Here's the bad stuff. I was doing a deliberate attack – a break-in battle but I did not have the necessary tools. I was doing a break-in battle with no assault engineers. There was no heavy assault engineer equipment in the country. We borrowed a civilian bulldozer and welded steel plate onto it.

All the stuff they taught us how to do in staff college came into play. If you don't have the resources, that's no excuse for not doing it, you go find it. I needed tanks, so phoned back to Canada, and they sent me tanks. I mean the walls of the fortified mud buildings over here are two feet thick. We need something to punch through.

Also, the roads were infested with mines and IEDs. We didn't have the necessary mine clearing equipment so I told Lieutenant-Colonel Lavoie, "go make your own roads to counter the IED threat." So we took a look at all the great vineyards out there and we said, have they harvested yet? Most people had harvested so we were not adversely effecting somebody's ability to make a living. We had to take a hard look at collateral damage as we went through the objective so as not to create more insurgents but the engineer shortage was a big factor of how we approached the problem. Another major problem was the national caveats in NATO. They are killing me, they are really killing me. We found out what NATO could not do. We simply couldn't get everyone we needed. The German wouldn't come down here. The French company weren't allowed to come down here. I couldn't get the Italians. We did get the Portuguese to come into the Kandahar Airfield to help out with static security tasks but most NATO countries came out with national caveats that precluded them from assisting us in actually fighting in Pashmul.

The idea of failing here [defeating the Taliban in Pashmul] was unacceptable. You want to talk about pressure, this was about a city [Kandahar City], a country, an alliance, and Canada was right in the middle of it, both from a battle group and from a brigade point of view. The battle was everything and failure was not an option. This was not just an attack; it was not just an operational fight; it was a NATO fight, it was everything and the more that we got into this fight, the more the pink cards – the un-stated national caveats started to creep into it. The more we got into the fight, the more we found that this was exactly what NATO was built for. This was almost Cold War-like type of fighting. It was conventional fighting. But not everyone was prepared to participate.

As the operation developed, it became more and more conventional. It was a conventional duke it out fight. The enemy wanted the ground and had prepared the area well for a defensive battle. In the end, it was all about putting the proper resources into the fight. We knew we would win because losing just was not an option.

So, when the enemy left we knew we had won this fight. However, we also realized that they would evolve. We knew the enemy would go back, they would go to ground for a bit [disperse and regroup in safe areas] and that they would do an after action review, after which they would come back at us in a far more sophisticated and dangerous way. They always do, they always adapt. The only question we had was how long was it going to take them to replace their leaders and how long was it going to take for them to come back at us again and what form would it take. When they did come back at us, they did so very quickly. They hit us with suicide attacks, IEDs and ambushes. So was it a surprise? No. Are they more dangerous now? Yes.

On a different note, reconstruction – its a lot harder than you can imagine. It's just a lot harder than any of the theory or even what we had experienced up until now. You go around that big table [interagency planning and coordination meetings] and everyone is arguing about their fights and I have won most meetings. I'd start laughing, and everyone would ask me what's so funny and I'd say, "I'm sorry but I've been looking around this room and honest to God, I am not Lawrence of Arabia but I sure feel like I'm sitting in that room [planning session for capture of Damascus from a scene from the film Lawrence of Arabia] – everyone's arguing and you have all the aid agencies saying we can't do this because your doing this and it's too dangerous, and we don't have enough information and I'd say listen to me we just won the biggest battle in NATO history and here we are sitting arguing about how we are going to rebuild stuff. I just started to laugh.

The battle was hard, but the reconstruction was harder. I've told people now, effects based operations, here's what it means to me, and I've defined what the team needs to fight full spectrum operations. Security operations measure effects in this environment in weeks. This battle will be felt for about 46 weeks effects on the ground.

The PRT [Provincial Reconstruction Team] measures its effects in months and years. We deliver food, which we're doing right now and it will help a family for eight weeks to three months. We build a road, which we are doing right now and we'll be measuring in years the effect of that road. Roads are huge. The OMLT [Operational Mentoring Liaison Team] with the Kandaks [ANA infantry battalions] out there measure their effects in generations. They are affecting generations and I told everybody around my table, I said, you have to come to me with your plans and answer that question across that temporal spectrum that I just described and how does it all tie together from weeks to generations. And, if you're missing any one part of that spectrum, or if there are gaps in that spectrum, it's not good enough to go back and complain – you have to give me possible solutions; you have to give me options of how to get that continuity of effect across that entire spectrum.

That is what we need to achieve because that's the effect that Canada is expecting from us for Afghanistan and here in Kandahar. Have you achieved that effect that is measured, in fact, in days to generations because there are some effects when you go out there, you only want to have a few days effect. And when you say it like that, then everyone knows how he fits into the big picture. It also helps drive where the decision points are when you have to have things up and running and delivering an effect.

When a security operation starts off, the PRT better be right there so it's kicking off in sync because you can't have a gap between the security operation and when the PRT starts delivering its effect because all of a sudden the conditions that the PRT needs are gone. The PRT operation has a complimentary and a compounding effect of reinforcing what that security operation achieved and taking it to the next level. Construction is security. It provides security in a more holistic and different way because it's going to enhance security in month timeframes and years.

Quite frankly security is a lot better in Pashmul now that the road is built because it enhances its freedom of movement. People are more confident now and they see something the Taliban cannot do. It also reduces the cost of food. That's what roads do. You can even put less soldiers on the road now. Now, you still have IEDs and ambushes but it's a lot better than it was before. In addition, you put police on the road. And how do you get a policeman out there? You have to train him, so someone trains him, now when he finishes his training, he's going to work for 20 - 30 years. So that effect complements PRT effect, which was complementing security effect. So you need everything working together to create that system.

So that's what this operation is doing – we finished it, it's gone, it's done. We are now moving on to the next operation. We are doing much more of the same – we're building, we're training capacity and we're carrying on with our security operations where the enemy is, which is in the southern part of the Arghandab River area.

The advice I would give to the next commander would be of a philosophical nature. We have trained all our commanders in a conventional symmetric manner but they are now operating in an unconventional environment. The latter requires flexibility, the former is a little bit more straightforward. I think when we train, we have to instill even more flexibility in our commanders. Despite what the previous focus has been, the Cold War days are gone. The basic warfighting idea, quite simply what and what TF 3-06 did during Operation Medusa works. But, what we need to arm our commanders with now is more complex issues, a faster tempo and complications that require mental flexibility to self-generate task forces inside of your formation; to regroup and manoeuvre and deal with constant change and manage your risk, but not become reckless.

This was a hard task for everybody involved. The requirement, particularly as a result of all of the national caveats, is the ability to generate forces, to get forces from elsewhere at the time when you needed them. The old way of thinking, the staff college solution, was if you don't get them, you don't go and attack. But that is not always an option – you have to learn how to generate the forces you require.

In fact, the DS [directing staff] at staff college would have said that I should not have done the attack. However, here's the thing, we couldn't fail here. The city was at risk, the country was at risk – we needed to win. So under these odds, under these conditions, I had to take the approach – go find me a solution. And, we found it.

What this tells me is that our training does work, but we need to modify our training to create even more flexible thinkers. One way is to make things even harder for people so they have to be innovative and think. In the end, this will make it easier for them to come to places like Afghanistan and when they come up against the next stronghold like in Pashmul, they'll be ready enough to do it.

Personally, my operational experience, working with the British, working with the Americans, working with the French, I was ready to do this. Now if I didn't have all that, I'm not sure if the normal staff college training would have been enough. Experience is key. In the end, it comes down to the right blend of experience, education and training. We need to ensure we prepare people properly and give them the right experience.

I had a good staff that supported me on operation and I had a lot of help from the Americans and just to make sure that, you know, I got what I needed here. Nonetheless, we've got to train our people to be more flexible and get outside the box but to do that in a sequential way. You've got to give them the foundation. They need to know and do all the conventional fighting, but also all the unconventional fighting so that we, again, in a situation like this, they know how to deal with both. They need to understand the risk and no matter how many corners you cut, if you don't give them a solid foundation, they'll never get to this stage. If someone told me I was going to do a conventional fight when I came over here, I would have just shaken my head ... I was shaking my head, I couldn't even believe I was asking Canada for tanks but I needed tanks. In the end, it was no small victory.

CHAPTER 14

WE THREE HUNDRED: LOGISTICS SUCCESS IN THE NEW SECURITY ENVIRONMENT

Lieutenant-Colonel John Conrad

Za canadai askar yam Za canadai askar yam --
"I am a Canadian Soldier"

Kandahar is the white-hot anvil upon which many timeless truths are being hammered out for the Canadian Forces (CF). Best practices that our veterans understood so well are being re-discovered on a daily basis in southern Afghanistan. Under the blistering test of combat operations, a generation of policy with bureaucratic emphasis is being stripped away and revised to better suit an army at war. In the case of my combat function, the fundamental requirement for rugged logistics to survive on the battlefield was reaffirmed. This brand of combat service support is not your grandfather's logistics but rather a new support capability that can withstand enemy attack and fight supplies and services through to the soldiers that need them. My experience serving with Operation Enduring Freedom in 2006 has left me with two inalienable truths: the contemporary battlefield – a complex operating environment with no discernable front or rear area – is lethal to ground-based logistics and the Canadian Army possesses a superb tactical logistics capability surprisingly well suited to operate upon it.[1] The latter capability has been forged more by good luck than serious interest from the Army's senior leadership. We were extremely fortunate with our first rotation into the south.

OPENING MOVES

I have been a soldier all of my adult life, heading off to basic officer training directly from high school in 1983. After taking my commission in the Canadian Navy in 1987, I transferred to the Transportation Corps, a combatant component of the unwieldy Logistics Branch, the ill-understood product of CF Unification. I have spent the bulk of my career in field units. I have commanded at every level inside the Service Battalion from platoon commander to Commanding Officer and have peacekeeping experience in Cambodia (1993) and Bosnia (2000). I was first warned of the Kandahar mission by way of a personal letter from Major-General Stu Beare in May of 2005. The letter explained that upon assuming command of 1 Service Battalion in the summer that I would generate and lead the logistics battalion, the National Support Element (NSE) in Kandahar in February of 2006.

I knew I was in for a pivotal challenge for Afghanistan is a country that has been smythed by God and man to impale armies and bleed them white. Everywhere one looks, the challenge to military sustainment resonates in Afghanistan. From the punishing alpine topography in the north to the searing white heat of the southern flood plain around Kandahar; from the paucity of good roads and well-articulated networks to the complete lack of railway, it is clear that the country is naturally inclined to deny the projection of combat power. This is exactly the way that generations of Afghans have wanted it. Upon assumption of 1 Service Battalion on 24 June 2005, I began to prepare the logistic architecture to sustain a mechanized Canadian battle group in full spectrum operations as well as a Canadian-led multinational brigade headquarters to replace the 173 (US) Airborne Brigade in southern Afghanistan. The unit would serve as the integral logistics company of the 1st Battalion, Princess Patricia Canadian Light Infantry (1 PPCLI), as well as the main sustainment platform for all Canadians in Afghanistan.

WE THREE HUNDRED

During the Battle of the Somme in 1916, the bottom dropped out of the sustainment system when the crushing demand for materiel for the offensive could not be met. Field Marshal Alexander Haig shocked the entire British Expeditionary Force (BEF) when he brought in a civilian expert to make it right. The problem Field Marshal Haig faced in 1916 was similar, albeit on a much larger scale, to the one that confronted our Canadian Task Force some ninety years later in Kandahar: the unknown. The pure weight of the Somme offensive in terms of the demand for fuel and ammunition far outstripped the ability of the BEF logistics apparatus to support it.

In the case of the 1 PPCLI Battle Group, it would be the first time since the Korean War that Canada's Army was deliberately putting itself in a sustained fight. Empirical Canadian calculus for consumption, repair and recovery in this sort of counter-insurgency operation did not exist. Unlike the Somme, our lines of communication could not take advantage of a large seaport like Boulogne and a sprawling inland railway. Rather, our logistics needs had to be prioritized through a narrow air bridge between Kandahar Airfield (KAF) and Camp Mirage – a dicey limitation with which to navigate unknown waters.

As one would expect, the construction and detailed organization of the NSE for Task Force 1-06 became my life. I threw my heart and soul into preparations, wrote the most detailed administrative appreciation I have ever rendered and applied widely for input from colleagues. After our initial Strategic Reconnaissance in August 2005, I passed my notes to friends and mentors to make sure I had not missed points. All sustainment personnel in the Canadian Forces are generalists save for one or two areas of specialization and I wanted to ensure that no vital ancillary services were undersubscribed because of my lack of a more detailed familiarity with their function. For example, Material Technicians are beyond my specialist area and I wanted to ensure that I had enough welders for the inevitable battle damage repairs.

By the time we headed back to Kandahar in October of 2005 for the Tactical Reconnaissance, I had a good outline plan that called for some additional support soldiers. The need for more troops was predicated by the requirement for combat projection in a much larger area of operation. Additionally, I was short of operational supply accounting capability. The NSE was going to be stocking over 32,000 line items for one mechanized battle group. A line item is like a species of animal; it does not refer to quantities but rather type. One line item may comprise a hundred individual units of that specific item and in truth we held millions of parts and widgets to keep Canada in the fight.

Hitler broke down in Russia in 1942 with nearly a million line items across three Army Groups.[2] Our level of item complexity was haunting in its similarity to the parts conundrum that doomed his Operation Barbarossa. It is not sexy to discuss the supply plumbing of a task force in our little army but the truth remains that forces fail if they cannot keep their supply house in order. I knew from the outset that we lacked sufficient supply specialists for the task. Adding new equipment and spare parts packages direct from industry as we did during the tour only exacerbated the manpower problem.

I was shocked early in the Tactical Reconnaissance when my boss came to see me on KAF and gave me a firm number for the size of the NSE – not one logistics soldier over three hundred! This was nearly the same strength of the Kabul-based NSE, which had been mostly static. Regional Command South was some 225,000 square kilometres of axle-snapping terrain by way of comparison. My unit was destined to project the sinews of combat over unheard of distances in one of the nastiest scenarios ever faced by Canadian logistics. Elsewhere on the Tactical Reconnaissance, good decisions were made to add an additional Light Armoured Vehicle (LAV) III company as well as an artillery battery, a Tactical Unmanned Aerial Vehicle (TUAV) flight and other specialist capabilities and equipment.

In all of this, the appetite to add commensurate service support troops was absent. I was reminded by senior Canadian officers on the brigade staff that the tail could not be permitted to wag the dog on an operation as important as this one. I understood the challenges faced by superiors – my boss's concern for soldier numbers and the Army Commander's pressures for long-term force generation. I do not believe, however, that they truly understood mine.

My stress levels hit the roof as we wrapped up pre-deployment training back in Western Canada. I worried constantly about logistics failing in Kandahar. I never doubted for a second that failure of Canadian logistics in Afghanistan would be mine to own notwithstanding any professional discussions that had gone before. With that in mind, I made two more unsuccessful overtures to the boss back in Canada, the last of which sought to add just three supply technicians to the NSE.[3] National staff officers from the former J4 Logistics Staff, challenged me incessantly. These senior logistics staff officers quite rightly challenged my math in terms of the correct numbers of troops for the tasks at hand in Kandahar.[4] I explained my predicament and they promised me support in the form of technical assistance visits (TAVs).[5]

For the first time in my career, I was grateful for National Staff assistance. I felt oddly disloyal but I took the Ottawa promises all the way to the bank. Now that national level logistic organizations had strengthened my operational posture, I decided in my framework Operation Order to stress tactical support. My main effort in Kandahar would be to support the moment to moment fight at all times and in all ways. I deemed I would always have a bit more time and dedicated national help to solve operational and strategic support problems. It was far from ideal but certainly the best hand to play after what we had been dealt.

IN THEATRE

The Regimental Sergeant Major (RSM) and I arrived in Theatre on 11 February near the front of the airflow. The brigade had wanted bayonets on the ground first but we had lobbied hard to get at least some of the soldiers of the NSE in country early. The NSE we were replacing did not have the same projection capability mine had and Combat Service Support (CSS) was needed immediately to support battle group mission rehearsal training.

The first issue that confronted us was a dramatic change in the agreed concept of battle group operations as worked out between Lieutenant-Colonel Ian Hope and myself in the fall of 2005. The initial concept of operations held that a maximum of two LAV III companies would operate simultaneously while the third one refitted on KAF. The LAV companies operating were expected to be conducting full spectrum operations for periods of five to seven days' duration.[6] The revised concept of operations called for all three LAV III companies to operate all of the time from forward operating bases. The change in concept of operations made eminent tactical sense but re-shaping the logistics battalion to support this dramatic manoeuvre change would be akin to steering RMS *Titanic* an hour after the iceberg collision. The small size of my unit would pinch more acutely with this shift in concept.

DANCES WITH NON-CONTIGUITY

The weight of enemy activity did not take long to affect my columns. Our convoys into Kandahar were ambushed on a number of occasions on their way to the Provincial Reconstruction Team (PRT) site, Camp Nathan Smith. We had prepared the soldiers for this kind of fighting but the realization that they were in it even before our Transfer of Authority (TOA) date of 1 March 2006 was somewhat disorienting. I recall one of my corporals going on his first convoy in late February and emptying nearly a complete magazine of 5.56mm ammunition into a machine gun position during a sophisticated ambush. The convoy escaped without human or vehicle casualties but my corporal was pretty charged up.

CHAPTER 14 IN HARM'S WAY

On 3 March 2006, two days after our contingent had taken over, a suicide attack on a combined Battle Group/NSE column resulted in a priority one casualty that had to be evacuated from the scene by one of my "Bison" armoured vehicle general purpose (AVGP) crews. We had confirmed by our TOA date what we had suspected all along – there can be no grey men or women in the unit.[7] On that day of days, it could be the female cook or the scrawny male radio repairman that applies the life saving action. The randomness of the battle demanded that all be prepared.

LEADERSHIP IN COMBAT

We had few helicopter sorties for logistic use and a relatively small unit for the support tasks at hand. With too few soldiers in our organization, it meant that some transport crews had to do upwards of four convoys a week. To mitigate these limitations, I decided that everyone regardless of rank, status or title in the NSE was obligated to serve on convoy duty. In this fashion, we could share the stress load on the dangerous spice trails that served as our main supply routes. Additionally, the RSM and I made it part of our battle rhythm to participate in one convoy a week. The nature of this battlefield is such that you can go from being the commanding officer to number two rifleman in the security cordon pretty quickly and whether you are a major or a corporal does not make much difference on the physical plane in an attack.

On the moral plane, however, participation of the chain of command was absolutely vital. The convoy experience is quite spiritual in the respect that the first move in a suicide or Improvised Explosive Device (IED) attack is really up to providence. The greatest fear that we shared was being killed without having the chance to fire a single shot. Whether you personally live or die depended so much on proximity, the armour of your truck and pure chance. If you live you will be participating in a section level fight and your chances of successful extraction at that point with our finely honed drills are excellent.

The leadership challenge centred on the randomness of the first move. I put a lot of energy into getting our fine soldiers past this impartial and shockingly violent opening assault. James Thurber once said something to the effect that, "A sense of humour is a serious thing." I absolutely agree. I encouraged all of my leaders from master-corporal to major to be as relaxed as possible in our environs. It seems trite writing it here now as

"THE BUCK STOPS HERE":

there were very few things in Kandahar operations that were funny. However I found humour to be a powerful elixir that enhanced rather than eroded our combat effectiveness. As a group, we joked and chatted with each other before the orders groups and at safe rest stops along the course of the day, but when it was time for business, there was always a palpable knife-edge on the convoy net. The RSM and I strove to look unfettered and happy all the time, in reality, this became ninety percent of my job.

REALITY HURTS

One of my most poignant memories from the tour comes from the end of our most successful operation in late July 2006. It serves most fittingly here as one example of an IED attack out of many our soldiers endured. The triumphant return of the Task Force would prove bittersweet for both the NSE and 1 PPCLI. The Task Force had operated together in our neighbour's province at distances sometimes exceeding 350 kilometres for over two weeks. The battle group had performed magnificently and destroyed a number of Taliban. The NSE, as the integral support unit for the Task Force was to establish a commodity point to top up the battle group in Helmand as they returned home.

On the morning of 22 July 2006, we left KAF at 0430 hours with the logistics vehicles to furnish the commodity point in Helmand. The column had two "Coyote" reconnaissance vehicle escorts as well as three "Bison" AVGPs from both the NSE and 1 PPCLI. This added up to a lot of escort firepower. On the western edge of Kandahar City, we were forced to halt to avoid a Taliban ambush that had engaged some Afghan National Security Forces (ANSF). A platoon from Major Grimshaw's "B" Company came to our aid and smoothed out the wrinkles in front of us.

We were on the move to Helmand Province again by 0600 hours. Two hours later we arrived at the planned location for the commodity point near the junction of Highway 1 and Secondary Road 611. The "Coyotes" and "Bisons" took up perimeter security and the CSS trucks were centrally positioned. The battle group began to arrive in platoon and company sized groups around 1100 hours and the last elements pulled in to the site around 1600 hours. After the last LAV III was filled with diesel, Lieutenant-Colonel Ian Hope gave a brief speech to the unit about the operation they had just completed. I spoke briefly with Ian before we got on the road. I had lost one of the "Coyote" escorts to mechanical break down and we agreed to stick close together in one large column as we made the long journey back to KAF.

Just before 1700 hours, I had to stop my logistics column to put a broken HLVW [heavy lift vehicle wheel] truck on the wrecker. After crossing the Arghandab River on Kandahar's western edge I had to make another brief stop to allow the mechanics to back off the brakes on the HLVW that was being towed. We were back on the move again by 1715 hours and sometime within the next 15 minutes we were hit by a vehicle borne IED.

The attack was typical yet surreal; a small Toyota Hiacre truck was slowly making its way along our convoy, travelling in the opposite direction on the north edge of the road. One moment I was dimly aware of the white Toyota cab and the next moment there was an enormous explosion and a large ball of smoke, body and vehicle parts expanding towards our diesel truck. For a second I thought the enemy had missed and we were about to accelerate and get the hell out of there. Then, in the next fraction of a second, we noticed the "Bison" that had been to my front before the fireball was in the ditch. We immediately set up a security cordon.

One of my other "Bisons" fired off a three round burst from the C6 general purpose machine gun (GPMG) at some suspicious looking bystanders who were observing the incident from the ditch with too much nonchalance. They fled across the open field to the north. On the south side of the road was a terraced village built into the side of a hill that afforded a spectacular view of our convoy. The "Coyote" escort called for a medical evacuation (medevac) and quick reaction force (QRF) assistance.

I checked in with the RSM and ran to the crippled "Bison" to assess the situation. Upon arrival at the "Bison" I noticed one of our men laid out on the ground behind the ramp where his comrades had been attempting first aid. His wounds were extremely serious and it was obvious that the soldier was dead. I later learned that this soldier was Corporal Warren of Montreal. Additionally the wounds to Corporal Gomez, the "Bison" driver were fatal and he had been killed instantly in the explosion. There were about four of us who worked on the priority one casualties in the "Bison." Captain Tony Ross from our S4 (Logistics) cell had been crew commanding the stricken Bison and he was throwing up from a head trauma. I held his hand and chatted to him.

IN HARM'S WAY　　　　　　　　　　　　　　　　CHAPTER 14

The interior of the Bison was a nightmare blend of dark blood, vomit, Red Bull cans and empty oatmeal cookie wrappers—bizarre juxtaposition of the mundane and the real. The QRF medic showed up from the PRT site some 35 minutes later. We extracted the casualties to the Black Hawk medevac helicopter and the put our deceased into the QRF "Bison." As the QRF went to leave the cordon, the second attack occurred.

The second attack had not penetrated our cordon but rather had waited for the QRF to exit the cordon with our dead. The suicide bomber in this case, was wearing a vest full of metal, ball bearings and the like over the explosive. There were a lot of Afghan casualties in the second explosion but no Canadian ones. Only one Mercedes G-Wagon belonging to Lieutenant Catton, the QRF Commander had been damaged.

By this time my head was swimming, my sense for the passage of time was warped but the light was starting to go down. I was certain that the Taliban were not done with us and I knew we needed to get moving before we lost the light entirely. I noticed a blue car streaking up into the terraced village after the second attack and there was no doubt in my mind that they were involved in the attack but they were moving too quickly in an area replete with civilians so I did not engage them. As well, all my wreckers were now full of vehicle casualties and if we could not tow Lieutenant Catton's G-Wagon with a light vehicle, I was contemplating destroying it in place.

An Afghan approached me and asked for help with their wounded. I directed him to the Afghan National Police (ANP) site leader. There were by now a number of civilian ambulances on the scene and our hands were full with our own problems.

It turned out that Catton's G-Wagon was ambulatory. I muttered a prayer of thanks as we dropped it onto a "Bison" A-Frame. At that very moment, two Apache Gunships appeared overhead like two angels of the Lord. They stayed over us while we slowly got underway and made our way through the liquid black streets of Kandahar to our PRT camp. By this time, the soldiers had been up for nearly 24 hours and I decided to stay there that night. We needed to re-group.

The action of 22 July was not dissimilar to many other days in the complex operating environment. Like so many incidents on our tour, the day was replete with shocking vignettes. It had been heart rendering to watch Afghan adults wrap up their children in blankets and scarves—the all too familiar practice of harvesting their dead. I will never escape the image of a civilian pick up truck with bodies being placed in it, blood streaming down the tailgate and over the bumper; the remains of the first suicide bomber being taken away in a Glad garbage bag. In the centre of it all, you grope for good solutions and push however clumsily to achieve what must be done.

ASPECTS ON THE MORAL PLANE

In Kandahar I learned more about Clausewitz and the war on the moral plane than I had accumulated in 24 years of service combined. So much of your success depends on mental toughness, the true backbone of manoeuvre. My first lesson was a heightened appreciation for all other trades. In peacetime exercises, one can be sceptical about the whiz kid communicator from the Signal Corps or the skinny female medic that looks as frail as onion skin. When you fight with these same soldiers doing their technical job under contact you grasp specialist raison d'être in a powerful new way. There was a tremendous sense in the contingent that everyone was your brother or sister and it was an invigorating combat multiplier, an enormous mental boost as you mounted up for a convoy.

NURTURING THE ELITE

Another distasteful and unfortunate phenomenon prominent on our tour was the whole inside the wire/outside the wire psychology. Certain aspects within the battle group cultivated this brand of elitism and it perplexed me as a senior leader in the task force. The "in the rear with the gear" mentality poorly applies in the contemporary operating environment. Everyone in that battlespace was subject to attack. KAF itself was rocketed over 45 times during the tour and one Canadian narrowly escaped death from a 107 mm rocket attack while eating supper at the KAF kitchen. Such is the nature of the lava lamp battlefield.

We lost soldiers from the medical corps and the NSE on the tour and I worked hard to shelter my unit from the "outside the wire" nonsense. It was particularly hard with the supply organization that was largely camp-based by necessity. Their efforts were vital to our success. They were every bit as crucial as my force projection pieces in the transport, repair and recovery platoons. Elitism has its useful corners but when nurtured in non-contiguity it has a corrosive effect. There were no bigger fans of the 1 PPCLI Battle Group than the NSE in Kandahar. I would never have tolerated a derisive word from my principal staff or non-commissioned officer (NCO) cadre that would portray 1 PPCLI in a negative light. I am not sure how well this particular two-way street was travelled.

IT IS NOT YOUR FAULT

I noticed among all of my leaders that they tended to blame themselves when their soldiers got hurt. What you must remind yourself and tell your subordinates is that these wounds of war are not your fault. You must ensure the dead are removed with the utmost respect and that the wounded are treated and are consistently well provided for until they return to duty or leave the theatre. If you have performed competently and fully discharged these duties you must move on. We, the living, need you. We need you fully invested in the next operation. Recognizing this emotion in myself and counselling my own non-commissioned members (NCMs) and officers in this aspect of mental toughness gave our leadership a certain Civil War earthiness.

This hands-on, grounded feel was exactly what I wanted in my command climate and I think it served us well. The soldiers needed to know that I knew them well. Formality and officious handling had no purchase in the NSE. All three hundred of us were in this fight together both inside and outside the wire.

THE FIRST DUTY

When you lose soldiers, a little piece of you dies along with him or her. That's certainly how I felt and feel. Corporals Gomez and Warren were lost in my convoy right in front of me on 22 July 2006. They were killed instantly and their ramp ceremony left me rocked by emotions I could never have anticipated. Similarly, when my outstanding Master-Corporal Raymond Arndt lost his life on escort duty to Spin Boldak on 5 August I took it hard. Ray was part of the young infantry militia platoon I had for convoy escort. Ian Hope had trained them to platoon level live fire and they came over to me for convoy escort training. They were fast learners and our investment in time, training and administration in that little platoon paid dividends for us throughout the tour.

CHAPTER 14

The RSM and I went down to the platoon lines to express our sense of loss and grief the night Ray died. The fine soldiers of that militia platoon were so young, so full of life and when they met my eyes with their collective grief it stirred me deeply. I was careful with my words to honour the memory of our comrade. I shook each of their hands after my brief chat and there were tears in my eyes when the RSM and I walked away.

I was counselled during my exit from Afghanistan for taking things too hard when my men were hurt or killed, but I absolutely believe that the leader has the first duty to remain human under these most inhumane of circumstances. You can show, in fact I believe you must, show emotion without losing control or respect. The soldiers will be disturbed if you behave otherwise.

I have been afraid many times in Kandahar this year. I can honestly tell you that the worst of my nightmares and the ugliest of my demons have not been of Taliban manufacture, but rather anchored in the fear of my tiny logistic unit running out of critical material or human capital. The CF achieved a great deal in 2006 and there is a tremendous amount for all of us who serve to be proud of. Our country and our tiny army are growing up around us in Kandahar. However, the consideration logistics derives from the leaders of the army is shockingly and inappropriately slim.

Logistics is more than ever a combatant partner in the battlespace, a crucible for tactical success. Yet, there are deeply rooted cultural biases in the CF that caused logistic services to be overlooked in Kandahar; that manifested itself in "fighting echelon" elitism that was truly incongruous in the contemporary operating environment. If we want to succeed in this battlespace you cannot push us away. We need each other to survive in this contemporary battle space. I know that we were extremely fortunate to succeed with the tiny CSS unit we sent to the fight in February 2006. I also know that my soldiers paid the price for this success. As proud as I am of the accomplishments of 1 PPCLI my heroes in Kandahar will always be those noble troops that lumbered north in 10 ton logistic trucks, "Bison" repair vehicles, aftermarket armoured wreckers and the like – no press, no glitter, just sheer guts.

ENDNOTES

1   The complex battlespace has been dubbed the non-contiguous battlefield by contemporary doctrine writers.

2   Martin Van Creveld, *Supplying War. Logistics from Wallenstein to Patton* (NY: Vail-Ballou Press, 1977), 151.

3   This final request for more support troops was tabled at an organization conference at 1 Canadian Mechanized Brigade Group Headquarters in Edmonton in December 2005.

4   This staff is now split between CEFCOM and CANOSCOM J4 staffs as part of CF Transformation.

5   This support was offered from Colonel J. Cousineau in early December of 2005. Cousineau was serving on the DCDS Staff as J4 Logistics at the time. A TAV represents additional specialist soldiers who deploy to a theatre for short durations to assist with the completion of a discreet task. The task is generally of such size and complexity that it is beyond the scope of the in place force to do it alone. I used supply TAVs to make basic material accounting possible in the theatre.

6   This was predicated on the average duration of most of the scraps that Task Force Gun Devil, the US Army predecessors to 1 PPCLI experienced in Kandahar Province during the previous year.

7   A "grey man" is a person who hangs back and relies on another stronger soldier to perform a sticky task whether it be to apply a tourniquet or send a 9-Liner Medevac request.

CONTRIBUTORS

Brigadier-General **D.A. Davies** is an experienced fighter pilot and leader who has flown the CF-5, F-16 and CF-18 aircraft. He commanded 425 Tactical Fighter Squadron, 3 Wing Bagotville, and Task Force Aviano during the first seven weeks of the ALLIED FORCE bombing campaign. His extensive staff experience includes NORAD HQ, USSPACECOM HQ, AIRCOM HQ, 1 Cdn Air Division, as well as NDHQ. He recently led the team that stood up CEFCOM, where he is currently serving as the Deputy Commander.

Lieutenant-Colonel **John Conrad** graduated from the Royal Military College of Canada (RMC) and was commissioned as a Naval Officer (MARS) in 1987. After retraining as an army combat logistics officer he served in virtually every command billet possible in 1 Service Battalion from platoon commander to Commanding Officer in 2005-2006. Most recently he has completed a tour as the Commanding Officer of the National Support Element, the logistics battalion responsible for sustaining the Canadian Task Force in Southern Afghanistan from February to August 2006. Lieutenant-Colonel Conrad is presently employed as a member of the Directing Staff at CFCSC.

Brigadier-General **David Fraser** is an experienced infantry commander who commanded at the battalion (2 PPCLI) and brigade (1 CMBG) level. He also commanded the NATO Multi-National Brigade in Sector South Afghanistan from February to November 2006. His operational experience includes tours in Cyprus, Bosnia-Herzegovina and Afghanistan. Brigadier-General Fraser holds a Master's in Defence Management and Policy from RMC and Queen's University. He was the recipient of the 2006 Vimy Award presented to the Canadian who made a significant and outstanding contribution to the security of Canada and to the preservation of our democratic values.

Rear-Admiral **Roger Girouard** began his Naval service at *HMCS CARLTON* in Ottawa. He has held a number of command positions including Commanding Officer of *HMCS MIRAMICHI* and *IROQUOIS*, Commander Maritime Operations Group Four in Esquimalt BC, Commander Canadian Fleet Pacific, and Commander Joint Task Force Pacific and Maritime Forces Pacific. He has deployed on a number of operations including OP FRICTION during the first Gulf War, OP PERSISTENCE - the CF element of the SWISSAIR 111 salvage and recovery operation, and he commanded the Joint Task Force for OP TOUCAN in East Timor, as well as the Canadian Naval Task Group for OP APOLLO in the Arabian Gulf.

Lieutenant-Colonel **Mike Goodspeed** is a serving officer in the PPCLI. He is currently employed in Kingston as a staff officer coordinating officer professional development in the Canadian Defence Academy. In addition to his service in South Sudan, Lieutenant-Colonel Goodspeed has served for a total of 26 years in the CF, as well as for eleven years in the private sector in both communications and high technology. He has served in all three battalions of his regiment as well as at the PPCLI Battle School and the Infantry School. He holds an MBA from the University of Calgary.

Lieutenant-Colonel **Ian Hope** (PPCLI) is currently serving as a liaison officer to United States Central Command in Tampa Florida. He commanded Task Force Orion in Kandahar from January to August 2006, under Operation ENDURING FREEDOM and later under ISAF IX. He has seventeen years of service in infantry battalions, including the Canadian Airborne Regiment and the British Parachute Regiment. He is a qualified strategic planner and has served two tours with US Combatant Command headquarters. He has a Bachelor of History (Honours) Degree from Acadia University, a Masters of Military Arts and Science from the United States Army Command and General Staff College; and is a graduate of United States Army School of Advanced Military Studies. He is currently working on his PhD dissertation in history with Queen's University.

Colonel, Dr. **Bernd Horn** is currently the Director of the Canadian Forces Leadership Institute. He is an experienced infantry officer with command experience at the unit and sub-unit level. He was the Commanding Officer of 1 RCR (2001-2003); the Officer Commanding 3 Commando, the Canadian Airborne Regiment (1993-1995); and the Officer Commanding "B" Company, 1 RCR (1992-1993). Colonel Horn holds an MA and PhD in War Studies from RMC where he is currently an Adjunct-Associate Professor of History.

Colonel **Mike Jorgensen** is currently the Commander of the Combat Training Centre in CFB Gagetown, New Brunswick. He is an experienced Infantry officer who has served in all three battalions of The Royal

Canadian Regiment, as well as the Canadian Airborne Regiment, and the Infantry School. He has commanded at company and battalion level, spending three years as Commanding Officer of The 3rd Battalion, The Royal Canadian Regiment. He is a graduate of the US Army Ranger Course, the CLFCSC, and the CFCSC. He has served in Germany, Cyprus, and the Former Republic of Yugoslavia, and he participated in major domestic operations such as Operation RECUPERATION (the ice storm), and the military response to the Red River Winnipeg flood. Recent senior staff appointments include Director of Army Training, and Director of CF Professional Development.

Colonel **Dean Milner** is an armour corps officer who served with The Royal Canadian Dragoons (RCD) at the troop, squadron and regimental level, commanding the RCD from 2002-04. He has a wide range of experience from operations and staff positions in the Joint Staff and Peacekeeping Policy Directorate. He deployed twice to Bosnia, the first time during the UNPROFOR days and the second time in 2003-04 as Battle Group Commander deployed in MNB (NW). He also deployed to Ethiopia/Eritrea as the Operations Officer for UNMEE. He is currently employed as Director Army Training responsible for professional development and all individual and collective training for the Army. He is a graduate of RMC and is currently working towards his Masters in War Studies.

Lieutenant-Colonel **G.L. Smith** is a C-130 Hercules pilot with extensive experience in the air mobility role. After multiple tours in training, operations, and staff roles over a twenty year period, his selection for command of the Airlift Detachment of OP APOLLO while Commanding Officer of 429 Squadron delivered an exhilarating challenge for which he ultimately received the Meritorious Service Medal in recognition of his efforts. A graduate of Royal Military College Kingston, Lieutenant-Colonel Smith has recently completed a tour within the Air Force staff at NDHQ as the first section head for unmanned aerial vehicle requirements.

Lieutenant-Colonel **Pierre St-Cyr** is an experienced tactical helicopter pilot with command experience from flight to squadron level, as well as on multiple UN and NATO missions such as Operations CONSTABLE (Haiti), PALLADIUM (Bosnia), HALO (Haiti), and ARGUS (Afghanistan). Following his actions in OP HALO, Lieutenant-Colonel St-Cyr received the Meritorious Service Medal. He has a Masters degree in Project Management from Université du Québec (Montréal). He is currently serving as the DND Liaison officer with the Canadian Space Agency and he is designated to become the special Liaison Officer for the CDS and CEFCOM Commander with Joint Force Command – Brunssum (Netherlands).

Roy Thomas is a retired Armour officer with UN service in Cyprus, the Golan Heights, Jerusalem, Afghanistan, Macedonia, Sarajevo and Haiti. Hi-jacked in South Lebanon and taken hostage in Bosnia, he is a recipient of the Meritorious Service Cross, an UNPROFOR Force Commander's Commendation, an UNMIH Force Commander's Commendation and a US Army Commendation Medal. A graduate of the Pakistan Army Command and Staff College, Quetta, Baluchistan, and the UK Tank Technology course in Bovington, Roy also holds an MA in War Studies from RMC.

Colonel **François Vertefeuille** is an experienced signals officer who had the opportunity to command a Brigade Signals Squadron and an infrastructure IM/IT organization within 5 Area Support Group. In his two command tours, he participated in OP ABACUS (passage to year 2000), OP RÉCUPÉRATION (1998 ice storm), and OP QUADRILLE (Sommet des Amériques in Québec City). He graduated from RMC in 1980 and is currently employed as the Commander Communication Reserve in NDHQ Ottawa.

GLOSSARY
OF ACRONYMS AND ABBREVIATIONS

3D	Diplomacy, defence and development
4GW	Fourth Generation Warfare
AAR	After Action Review
ACC	Air Component Commander
ADF	Australian Defence Force
AFV	Armoured Fighting Vehicles
AJAG	Assistant Judge Advocate General
ALCE	Airlift Control Element
ALTF	Airlift Task Force
AMIB	Allied Military Intelligence Battalion
ANA	Afghan National Army
ANP	Afghan National Police
ANSF	Afghan National Security Forces
AO	Area of Operations
AOR	Area of Responsibility or Auxiliary Oil Refueller
APC	Armoured Personnel Carriers
AVGP	Armoured Vehicle General Purpose
Bde	Brigade
BEF	British Expeditionary Force
BG	Battle Group
BiH	Bosnia-Herzegovina
BOI	Board of Inquiry
C3I	Command, Control, Communications and Intelligence
CA	Civil Affairs
CAST	Canadian Air Sea Transportable
CCAT	Contingent Commander's Advisory Team
CCO	Combat Control Officer
CDS	Chief of the Defence Staff
CEFCOM	Canadian Expeditionary Command
CENTCOM	Central Command
CEU	Construction Engineer Unit

GLOSSARY

CF	Canadian Forces
CFB	Canadian Forces Base
CFC-A	Combined Force Command Afghanistan (CFC-A)
CFE	Canadian Forces Europe
CIDA	Canadian International Development Agency
CIMIC	Civil-Military Cooperation
CJ	Combined Joint
CJTF	Canadian Joint Task Force
CLS	Combined Land Staff
CMBG	Canadian Mechanized Brigade Group
CMO	Chief Military Observer
CP	Command Post
CPA	Comprehensive Peace Agreement
CO	Commanding Officer
COG	Centre of Gravity
COP	Covert Observation Platoon
COS	Chief of Staff
COY	Company
CSS	Combat Service Support
CTF	Combined Task Force
CPX	Command Post Exercise
DCDS	Deputy Chief of the Defence Staff
DFAIT	Department of Foreign Affairs and International Trade
DFID	Department For International Development
DIN	Defence Information Network
DND	Department of National Defence
DPKO	Directorate of Peacekeeping Operations
DPRE	Displaced Persons, Refugees and Evacuees
DS	Directing Staff
EAF	Entity Armed Forces
ERT	Emergency Response Teams
EU	European Union
EW	Electronic Warfare
Ex	Exercise
FAC	Foreign Affairs Canada
FLIR	forward looking intra-red
FOB	Forward Operating Base

GFAP	General Framework Agreement for Peace
GOA	Government of Afghanistan
GPMG	General Purpose Machine Gun
GPS	Global Positioning System
Hel Det	Helicopter Detachment
HQ	Headquarters
HLTA	Home Leave Travel Allowance
HLVW	Heavy lift vehicle wheeled
HMAS	Her Majesty's Australian Ship
HMCS	Her Majesty's Canadian Ship
HSS	Health Service Support
HVT	High Value Target
ICOM	communication intercept
ICTY	International Criminal Tribunal for the former republic of Yugoslavia
IDP	internally displaced persons
IED	Improvised Explosive Device
IFF	Identification Friend or Foe
IO	Information Operations
INTERFET	International Force East Timor
IPTF	International Police Task Force
IR	information requirements
ISAF	International Security Assistance Force
ISTAR	Intelligence Surveillance Targeting Acquisition Reconnaissance
JAG	Judge Advocate Generals
JCO	Joint Command Observer teams
JMCO	Joint Monitoring and Control Organization
JTF	Joint Task Force
JTF 2	Joint Task Force 2
KAF	Kandahar Airfield
Km	kilometre
LAOC	Law of Armed Conflict
LAV III	Light Armoured Vehicle III
LO	Liaison Officer

GLOSSARY

LOTS	Liaison Observation Teams
LRP	Long Range Patrol
MARLANT	Maritime Atlantic Command
MARPAC	Maritime Pacific Command
MARS	Maritime Surface Officer
Medevac	Medical Evacuation
MIF	Multinational Interim Force
MNB	Multi-National Brigade
MND	Minister of National Defence
MND SW	Multi-National Division South West
MOST	Monitoring, Observation, Surveillance Teams
MP	Military Police
NAI	Named Areas of Interest
NATO	North Atlantic Treaty Organization
NCC	Naval Component Commander
NCE	National Command Element
NCM	Non-Commissioned Member
NCO	Non-Commissioned Officers
NDCC	National Defence Coordination Centre
NDHQ	National Defence Headquarters
NDIC	National Defence Intelligence Centre
NGO	Non-Governmental Organization
NIPRNet	National Internet Protocol Router Network
NORCOM	Northern Command
NSE	National Support Element
NSU	National Support Unit
OC	Officer Commanding
O Gp	Orders Group
OHR	Office of the High Representative
OMLT	Operational Mentoring Liaison Team
OP	Observation Post or Operation (based on context)
OPCOM	Operational Command
OPCON	Operational Control
OPFOR	Opposition Force
OSCE	Organization for Security and Cooperation in Europe
PCC	Provincial Coordination Centre

PDC	Provincial Development Committee
PDSS	Personnel Designated Special Status
PER	Personnel Evaluation Report
PIFWC	Persons Indicted for War Crimes
PIR	Priority Information Requirements
POL	Petroleum, Oil and Lubricants
POLAD	Political Advisor
PPCLI	Princess Patricia's Canadian Light Infantry
PPE	Personal protective equipment
PRT	Provincial Reconstruction Team
PT	Physical Training
PX	Post Exchange
OGD	Other Government Departments
QRF	Quick Reaction Force
ORO	Operations Room Officer
R22eR	Royal 22nd Regiment
R&R	Rest and Recreation
RAAF	Royal Australian Air Force
RC-S	Regional Command-South
RCD	Royal Canadian Dragoons
RCMP	Royal Canadian Mounted Police
Recce	Reconnaissance
RISTA	Reconnaissance Intelligence Surveillance and Target Acquisition
ROE	Rules of Engagement
RO-RO	Roll on – Roll off
ROTO	Rotation
RPG	Rocket Propelled Grenade
RRB	Radio Re-Broadcast
RSM	Regimental Sergeant Major
SACEUR	Supreme Allied Commander Europe
SAF	Sudanese Armed Forces
SAR	Search and Rescue
SATCOM	Satellite Communications
SFOR	Stabilization Force
SITREP	Situation Report
Sqn	squadron

SMO	Senior Military Observer
SOF	Special Operations Forces
SOFA	Standing of Forces Agreement
SPLA	Sudanese People's Liberation Army
SRSG	Special Representative to the Secretary-General
Tac Avn	Tactical Aviation
Tac Hel Sqn	Tactical Helicopter Squadrons
TAV	Technical Assistance Visit
TEZ	Total Exclusion Zone
TF	Task Force
TFA	Task Force Aviano or Afghanistan (dependent on context)
TFH	Task Force Haiti
TIC	Troops in Contact
TNI	Indonesian Army
TO&E	Table of Organization and Equipment
TOA	Transfer of Authority
TOW	Tube Launched, Optically Sited, Wire Guided
TTP	Tactics, techniques and procedures
TUA	TOW Under Armour
TUAV	Tactical Unmanned Aerial Vehicle
UN	United Nations
UNAMA	United Nations Assistance Mission in Afghanistan
UNHCR	United Nations High Commissioner for Refugees
UNMIS	United Nation's Mission in Sudan
UNMO	United Nations Military Observer
UNPROFOR	United Nations Protection Force
UNTAET	United Nations Transitional Administration East Timor
US	United States
UXO	Unexploded Ordnance
VER	Vertical Ejector Racks
VIP	Very Important Person
WBS	work breakdown structure
WFP	World Food Program
WG	Working Group

INDEX

Afghanistan i, ii, 15, **22** *notes*, 24 *notes*, 110, 112, 113, 120, 121, 127, 135, 176, 179, 180, 185-187, 189, 191, 195, 196, 198, 199, 202, 203, 206, 208, **209** *notes*, 211-215, 225-228, 233, 235, 245, 254, 256-258, 260, 268, 271, 273, 274, **276** *gloss.*, **277** *gloss.*, **280** *gloss.*

Afghan National Army (ANA) 191, 199, 215, 229, 230, 232-234, 236, 243, 246, 254, **275** *gloss.*

Afghan National Security Forces (ANSF) 198, 199, 204, 214-216, 219, 233, 241, 263, **275** *gloss.*

Aircraft
  A-10 Thunderbolt 243, 247
  Apache Gunships 265
  C-130 Hercules 29, 109, 110, 112, 115, 116, 273
  CF-18 Hornet 89, 271
  CH-146 Griffon 139, 153
  Sea King Helicopter 45
  Seaknight Helicopter 42

Airlift Task Force (ALTF) 31, 33, 36, 38, 40, **275** *gloss.*

Alexandre, Boniface 138

Arghandab River 248, 250, 255, 264

Aristide 138

Assistant Judge Advocate General (AJAG) 104, 105, **275** *gloss.*

Australia 29, 31, 32, 36

Australian Defence Force (ADF) 31, 33, 34, 36, **275** *gloss.*

Aviano 89, 97, 271, **280** *gloss.*

Banja Luka 78

Bihac 58, 61, 63, 65-67, 75

Bonser, Commodore Mark 33

Bos Petrovac 58, 70

Bosnia-Herzegovina (BiH) (See also Bosnia) 3, 5, 73, 77, 83, 85, 87

Bosnian Croats 2, 3

Bosnian Serbs 1-3, 10, 12, 15-17, **21** *notes*, **23** *notes*

C3I 30, **275** *gloss.*

Camp Maple Leaf 41, 58

Canadian Engineer Regiment (CER) 58

Canadian Expeditionary Command (CEFCOM) 94, **269** *notes*, 271, 274, **275** *gloss.*

Canadian Forces (CF) iii, v-vii, **viii** *notes*, 2, 18, 19, 21, 27, 30, 38, 41, 50, 61, 68, 90, 94, 137-142, 151, 170, 171, **177** *notes*, 179, 181, 184, 189, 206, 207, 222, 226, 227, 257, 259, 268, **269** *notes*, 272, **276** *gloss.*

Canadian Forces Base (CFB) 50, 53, 272, **276** *gloss.*
  Petawawa 48, 50, 51, 72, 73
  Trenton 54, 109, 118, 119, 121
  Valcartier 42, 140

Canadian Force Personnel Support Agency (CFPSA) 119

Canadian International Development Agency (CIDA) 59, 64, 197, 208, 215, **276** *gloss.*

Caribinieri 65-67

Chief of the Defence Staff (CDS) 31, 38, 48, 61, 68, 72, 94, 99, 110, 118, 142, 179, 189, 212, 274, **275** *gloss.*

Civil Military Cooperation (CIMIC) 57-59, 64, 69, 145, 187, 201, **276** *gloss.*

CNN 6, **22** *notes*, **209** *notes*

Combat Service Support (CSS) 56, 257, 261, 263, 268, **276** *gloss.*

INDEX

Combined Force Command Afghanistan (CFC-A) i, 179, 180, 182, 183, 185, 186, **276** *gloss.*
Combined Task Force (CTF) Aegis ii, 243, **276** *gloss.*
Comprehensive Peace Agreement (CPA) 164, 167, **275** *gloss.*
Contingent Commander's Advisory Team (CCAT) 60, 62, **275** *gloss.*
Coralici 58, 59, 70, 79
Cosgrove, Major-General Peter 33-35, 43
Counter-insurgency 14, 198, 199, 205, 207, 214, 220, 226, 231-234, 236, 244
Courage v, 7, 10-13, 176, 206, 218, 219
Covert Observation Platoon (COP) 63, 75, **276** *gloss.*

Dallaire, Major-General Roméo 132
Darfur 167
Darwin 32, 33, 36, 37, 39, 40, 43
Delves, Major-General Cedric 54, 65, 66
Department For International Development (DFID) 59, **276** *gloss.*
Department of Foreign Affairs and International Trade (see DFAIT and see also FAC) **276** *gloss.*
Deputy Chief of the Defence Staff (DCDS) 31, 38, 48, 53, 54, 61, 94, 95, 179, **269** *notes*
DFAIT **276** *gloss.*
Dili 28-30, 32-39, 43
Directorate of Peacekeeping Operations (DPKO) 168, 169, **276** *gloss.*
Drvar 58, 63, 68, 70, 75

East Timor 27-30, 32-34, 36, 37, 39, 40, 42-45, 139, 173, 272, **277** *gloss.,* **280** *gloss.*
Exercise Storming Bear 52

Fear 36, 84, 89, 175, 187, 195, 196, 199, 218, 223, 224, 226, 262, 268

Foreign Affairs Canada (FAC) See also DFAIT 31, 197, 208, 215, **276** *gloss.*
Former Republic of Yugoslavia 55, 89, 103, 273
Fort Chaffee 51
Fourth generation warfare (see 4GW) 195, **208** *notes,* **209** *notes,* **275** *gloss.*
Fraser, Brigadier-General David ii, **23** *notes,* 243, 244, 271

Garang, John 170
General Framework Agreement for Peace (GFAP) 47, 55-57, **277** *gloss.*
Gonaïves 138
Gorazde 1, 4, 5, 15, 16, **22** *notes*
Grbavica 3

Haiti 30, 41, 137-140, 144-146, 148, 149, 157, 160, 273, 274, **280** *gloss.*
Helmand 221, 230, 248, 263
Henault, General Raymond 118
Hillier, General Rick 47, 49-51, 212
HMAS *Jervis Bay* 39
HMCS *Protecteur* 30-32, 34, 36, 38, 40-42
Hope, Lieutenant-Colonel Ian ii, 211, 244, 261, 263, 267, 272

IED 191-193, 195, 196, 217, 219, 221, 233, 237, 244, 252, 253, 254, 262-264, **277** *gloss.*
Improvised Explosive Device (See IED) 192, 217, 233, 244, 262, **277** *gloss.*
Indonesia 27, 29, 34, 43
Indonesian Army (TNI) 28-30, 34-36, 43, **280** *gloss.*
Information Operations (IO) 59, 77, 83-85, 219-221, 232, **277** *gloss.*
International Criminal Tribunal for the Former Republic of Yugoslavia (ICTY) **21** *notes,* **24** *notes,* **25** *notes,* 55

International Force East Timor (INTERFET) 27, 29, 31-40, 44, **277** *gloss.*
International Police Task Force (IPTF) 55, 59, 64, 67, 76, **277** *gloss.*
International Security Assistance Force (ISAF) 185, 187, 188, 206, 212, 213, 235, 236, 243, 245, 249, 251, 272, **277** *gloss.*
Interpreters 7, 9, 10, **23** *notes*, **24** *notes*, 198
Italian Caribinieri (see Caribinieri) 65

Joint Command Observer (JCO) teams 60
Joint Monitoring and Control Organization (JMCO) 165, **277** *gloss.*
J-Staff 52, 53
Joint Staff (see J-Staff) 273
Joint Task Force 30, 271, 272, **276** *gloss.*, **277** *gloss.*
Joint Task Force 2 (JTF 2) 69, **277** *gloss.*

Kabul 18, **24** *notes*, 179, 193, 213, 260
Kandahar 191, 201, 206, 211-213, 216, 222, 227, 228, 230, 231, 236, 243-245, 248, 251, 252, 254, 257-261, 263-268, **269** *notes*, 272, **277** *gloss.*
Kandahar Airfield (KAF) 191, 201, 222, 231, 243, 252, 259-161, 263, 264, 266, **277** *gloss.*
Kandahar City 191, 237, 243, 252, 263
Khartoum 164, 173
Kiseljak 2
Knin 69, 70

Land Forces Central Area (LFCA) 48, 53, 61
LAV III 191, 192, 194, 238, 261, 263, **277** *gloss.*
Lavoie, Lieutenant-Colonel Omer ii, 227-241, 247, 250, 252
Law of Armed Conflict 102, 103, 201, 222, **277** *gloss.*

Leadership i-iii, v-viii, 7, 12, 14, 15, 19, 20, **25** *notes*, 33, 34, 47, 48, 52, 58, 62, 63, 80-82, 89, 91, 99, 100, 105, 109, 115, 117, 118, 122-135, 137-139, 141-146, 148-155, 157, 159-161, **162** *notes*, 163-165, 167, 169-172, 174-176, **177** *notes*, 181, 183-185, 191, 195-200, 202, 206-208, 213, 214, 221, 224-226, 228, 231-234, 241, 247, 257, 262, 267, 272
Leclerc, Chantale 156, **162** *notes*

Ma'Sūm Ghar 229, 237
Mackenzie, Major-General Lewis **24** *notes*, 112
Martin Brod 67-69, 72
Monitoring, Observation, Surveillance Teams (MOST) 85, **278** *gloss.*
Mount Igman 1, 10, **21** *notes*
Multi-National Brigade (MNB) 85, 86, 243, 244, 271, 273, **278** *gloss.*
Multi-National Division South West (MND SW) 54, **278** *gloss.*
Multinational Interim Force (MIF) 139, 140, **278** *gloss.*

National Defence Coordination Centre (NDCC) 53, **278** *gloss.*
National Defence Headquarters (see NDHQ) 19, 36, 47, 120, 179, **278** *gloss.*
National Defence Intelligence Centre (NDIC) 53, **278** *gloss.*
National Support Element (NSE) 38, 44, 51, 70, 76, 86, 258-263, 266, 267, 271, **278** *gloss.*
NATO 2, 15, 17, **22** *notes*, **25** *notes*, 43, 61, 74, 76, 84, 86, 93, 105-107, 116, 174, 176, 187, 206, 207, 212, 213, 229, 230, 235, 237, 238, 243, 244, 251-253, 271, 273, **278** *gloss.*
NDHQ 19, 31, 36, 47, 120, 179, 271, 273, 274, **278** *gloss.*

NGOs 64, 74, 173, 197
New Zealand Army 36
Non-governmental Organizations (see NGOs) 173
North Atlantic Treaty Organization (see NATO) **278** *gloss.*

Office of the High Representative (OHR) 55, **278** *gloss.*
Operation ABACUS 27, 29, 274
Operation APOLLO i, 109, 110
Operation CAVALIER 83
Operation Enduring Freedom 189, 212, 257, 272
Operation HALO i, 137-139, 141, 144, 154, 160, 273
Operation MEDUSA ii, 228, 230, 235, 236, 238, 239, 243, 244, 255
Operation PALLADIUM i, 47, 83, 273
Operation RECUPERATION 49, 50, 273
Operation SHANNON 47, 67
Operation TOUCAN i, 27, 30, 31, 36, 41, 43-45, 139, 272
Operational Mentoring Liaison Team (OMLT) 254, **278** *gloss.*
Organization for Security and Cooperation in Europe (OSCE) 55, 59, 64, **278** *gloss.*

Panjwayi 221, 226, 236, 243, 244, 246, 247, 250
Pashtunwali 198
Pashmul 237, 240, 244-247, 252, 254, 256
Persons Indicted for War Crimes (PIFWC) 62, 63, **279** *gloss.*
Petawawa 48, 50, 51, 72, 73
Physical Fitness 96
Port-au-Prince 137, 147, 153
Princess Patricia's Canadian Light Infantry (PPCLI) 120, 211, 212, 214, 258, 271,

272, **279** *gloss.*
Provincial Coordination Centre (PCC) 214, **278** *gloss.*
Provincial Development Committee (PDC) 214, **279** *gloss.*
Provincial Reconstruction Teams (PRT) 179, 185-188, 211, 214, 215, 254, 255, 261, 265

RISTA 59, 279 gloss.
Rose, General Sir Michael 3, 8, 15, 21 *notes*, **22** *notes*, **24** *notes*
Royal 22nd Regiment (R22eR) 36, 40, 41, 44, 45, **279** *gloss.*
Royal Canadian Dragoons (RCD) 49, 50, 57, 58, 83, 84, **279** *gloss.*
Royal Canadian Mounted Police (RCMP) 45, 197, 273, **279** *gloss.*
Royal Canadian Regiment (RCR) 45, 197

Sarajevo i, 1-7, 10-21, **22-25** *notes*, 274
SFOR 54, 55, 61, 65-68, 72, 74, 76, 77, 84-86, **279** *gloss.*
Siah Choy 248-250
Somalia Commission of Inquiry 142, **161** *notes*
Soubirou, General 2, 10, 15, **23** *notes*
Special Representative to the Secretary-General (SRSG) 166, 167, **177** *notes*, **280** *gloss.*
Sperwan 246, 248-250
Spin Boldak 267
Stabilization Force (see SFOR) 54, 139, **279** *gloss.*
Status Of Forces Agreement (SOFA) 74, **280** *gloss.*
Suai 35, 39, 44
Sudan i, 163-167, 170, 172, 173
Sudanese Armed Forces (SAF) 164, 165, **279** *gloss.*

Sudanese People's Liberation Army (SPLA) 164, 165, 169, **280** *gloss.*

Tactical airlift detachment 109, 111, 113-116, 125, 126, 133
Taliban 112, 121, 197, 199, 201, 204, **209** *notes*, 213, 215-217, 219-221, 230-233, 237-241, 243-252, 254, 263, 265, 268
Task Force 1-06 (TF 1-06) 259
Task Force 3-06 (TF 3-06) 227, 248-250
Task Force 31 246, 248-250
Task Force Aviano 89, 271, **280** *gloss.*
Task Force Afghanistan 191, 271, **280** *gloss.*
Task Force Grizzly 248, 250, 251
Task Force to Haiti (TFH) 139, 140, **280** *gloss.*
Task Force Orion ii, 211, 212, 215, 219-221, 224, 244, 272
Three Block War 200, **209** *notes*
Trenton 54, 109, 110, 116, 118, 119, 121, 126, 127

Ulmer, Lieutenant-General W. E. 143, **161** *notes*
Una River 68-71
United Nations (UN) i, 1-5, 7, 10, 11, 14, 16, 18, **21** *notes*, **22** *notes*, 27-29, 35, 43-45, 55, 64, 74, 83, 138-140, 163, 165-176, 186, 207, 273, 274, **280** *gloss.*
United Nations charter 139, 165
  Chapter VI 165, 167, 169, 171, 172, 176
  Chapter VII 139, 165
United Nations Assistance Mission in Afghanistan (UNAMA) 186, 188, **280** *gloss.*
UN civil police 64
United Nations High Commissioner for Refugees (UNHCR) 17, **24** *notes*, 55, 59, 64, **280** *gloss.*
United Nations Military Observer (UNMO) 1-21, **21-25** *notes,* 173, **177** *notes*, **280** *gloss.*
United Nation's Mission in Sudan (UNMIS) 163-167, 171-175, **177** *notes*, **280** *gloss.*
United Nations Protection Force (UNPROFOR) 1, 2, 5, 8, 10, 11, 13, 19, **21** *notes*, **22** *notes*, **24** *notes*, **25** *notes*, 83, 273, 274, **280** *gloss.*
United Nations Security Council 139, 140, 166
Resolution 1529 139
United Nations Transitional Administration East Timor (UNTAET) 27, 44, 45, **280** *gloss.*

War on Terrorism 110, 171
Watt, Major-General Rettie 70
World Food Program (WFP) 173, **280** *gloss.*

Yugoslavia (see Former Republic of Yugoslavia) 19, **21** *notes*, 55, 89, 103, 116, 119, 273, **277** *gloss.*

Zagreb 5, 10, 11, 15, **23** *notes*, 54
Zepa 2, 5, 16, 17, **24** *notes*
Zgon 58, 62, 70, 77, 80
Zumalai 38, 40, 41, 44, 45

1st Battalion, Princess Patricia's Canadian Light Infantry (1 PPCLI) 211, 212, 214, 258, 263, 267, 268, **269** *notes*
1st Battalion, Royal Canadian Regiment (1 RCR) 48, 50, 52, 54, 78, 227, 246, 248, 272
1st Battalion, Royal New Zealand Infantry Regiment (1 RNZIR) 36

INDEX

1 Construction Engineering Unit (1 CEU) 40, 275 gloss.
1 Guards (Croatia) Brigade 63, 69
2nd Battalion, Royal Canadian Regiment (2 RCR) 57, 72
2nd Canadian Mechanized Brigade Group (2 CMBG) 47-50, 52
3rd Battalion, The Royal Canadian Regiment (3 RCR) i, 47-50, 54, 55, 57, 58, 62, 68, 73, 273
3D - Diplomacy, defence and development 200, 208, 215, **275** *gloss.*
4GW (see fourth generation warfare) 200, 202, 205, **208** *notes,* **209** *notes,* **275** *gloss.*
8 Wing Trenton 110, 116, 126, 127
11 September 2001 109, 110, 118
14 Wing 126, 127
430 Tactical Helicopter Squadron 137, **279** *gloss.*
438 Tactical Helicopter Squadron 137, **279** *gloss.*